D0218097

Anatomy and Physiology: A Guided Inquiry

Patrick J.P. Brown, Ph.D
ETSU College of Public Health

POGIL
WWW.POGIL.ORG
Copyright © 2015

POGIL Project
Director: Richard Moog
Associate Director: Marcy Dubroff
Publication Liaison: Sarah Rathmell

Copyright © 2016 by The POGIL Project. All rights reserved.

No part of this publication may be reproduced, stored in a retrieval system, or transmitted in any form or by any means, electronic, mechanical, photocopying, recording, scanning, or otherwise, except as permitted under Section 107 or 108 of the 1976 United States Copyright Act, without either the prior written permission of the Publisher, or authorization through payment of the appropriate per-copy fee to the Copyright Clearance Center, Inc., 222 Rosewood Drive, Danvers, MA 01923, (978) 750-8400, fax (978) 750-4470, or on the web at www.copyright.com. Requests to the Publisher for permission should be addressed to The POGIL Project, 713 College Avenue, PO Box 3003, Lancaster, PA 17604-3003.

ISBN-13: 978-1-119-17525-4
ISBN-10: 1-119-17525-9
V10015184_102819

Dedication

For Stacy

You ignited my passion for putting the student in command of their own learning and you continue to inspire me to this day.

About the Author

Dr. Patrick Brown teaches undergraduate Health Sciences at East Tennessee State University, primarily introductory anatomy and physiology. Additionally, he teaches a writing-intensive clinical parasitology course and co-teaches a graduate course for faculty entitled "Teaching for Learning in Higher Education". Patrick received his B.S. degree from the University of Tennessee at Chattanooga in Biological Science and his Ph.D. from the University of Georgia in Cellular Biology.

Patrick was introduced to POGIL by his wife, Dr. Stacy Brown, and after attending a three-day workshop in 2008 immediately began writing activities for introductory anatomy and physiology. Since then he has gone on to not only author this collection of activities, but has become a workshop facilitator and served as the Southeastern Regional Coordinator for the POGIL project. He has led seminars and workshops in active learning throughout the United States.

Acknowledgements

Special thanks are due to Rick Moog and Marcy Dubroff for making this whole endeavor possible. Their guidance and leadership of the POGIL Project has improved access to evidence-based teaching for thousands of students and instructors throughout the world. I also wish to thank my mentors in the Southeastern Regional POGIL network for introducing me to the POGIL method and training me as a workshop facilitator, particularly Dr. Andrew Bressette, Dr. Sally Hunnicutt, Dr. Suzanne Ruder, Dr. Gail Webster and Dr. Rob Whitnell. Thanks are also due to Emily Mavridoglou for her excellent copy-editing and formatting. Finally I have to thank my family for their constancy and support over the years – I couldn't do it without them.

Description of Roles for POGIL Activities

Manager
Manages the group. It's your job to ensure that everyone is doing their job, and only their job. You are also responsible for making sure that everyone participates and understands the concepts. The only person who is allowed to speak to the Instructor is the Manager.

Spokesperson
Presents group answers to the class. If the Instructor asks for an oral or written (on the board) response from the group, it will be provided by the Spokesperson. NO ONE ELSE is allowed to add or qualify what the Spokesperson says. If you want to make sure that your understanding is presented, the time to do it is during group discussions.

Scribe
Provides the official record for the group. The Spokesperson is the one who shares the group answers with the class, but the Scribe's answers are the official written record for the group. If the Scribe's answer is wrong, the entire group could suffer.

Librarian
Performs all technical operations for the group. The Librarian/Technician is the fact-checker and is the only person in the group who is allowed to have their book out, and then only if expressly allowed by the activity or the Instructor. The Librarian is also the keeper of the "clicker". When the Instructor asks for responses via the turning-point clickers, the Librarian will actually enter the group's answer.

Reflector
Provides insight and feedback regarding group dynamics. Are the members of the group fulfilling their roles properly? Is everyone afforded an opportunity to be heard? Is anyone monopolizing the conversation? These are questions the reflector should be able to answer if called upon by the Facilitator.

Consensus Builder
Steers the group towards consensus. If there is disagreement between group members, it is the job of the Consensus Builder to seek common ground and resolve disputes. The Consensus Builder should seek to ensure that he/she understands what group members are trying to communicate in an effort to find agreement.

POGIL
WWW.POGIL.ORG
Copyright © 2015

Table of Contents

Activity	Topic	Page

Introduction to Anatomy and Physiology

Activity 1	The Language of Science and Medicine	9
Activity 2	The Building Blocks of Matter	15
Activity 3	Acids, Bases, and pH (Oh My!)	21
Activity 4	Cell Transport Mechanisms	25

Histology and Integument

Activity 1	Basic Histology	33
Activity 2	Skin and Temperature Control	37

Musculoskeletal System

Activity 1	Range of Motion	43
Activity 2	Ossification	49
Activity 3	Articulations	55
Activity 4	The Sliding Filament Theory	61

General Nervous

Activity 1	Membrane Potentials	69
Activity 2	Conduction of Action Potentials and Synapses	77
Activity 3	Reflex Arcs - The Simplest Neural Circuit	83
Activity 4	Receptors, Receptors, Receptors	87

Endocrine

Activity 1	Endocrine Glands and Hormones	93
Activity 2	Hormone Mechanism of Action	99
Activity 3	Regulation of Endocrine Secretion	105

Blood

Activity 1	Red Blood Cells	113
Activity 2	ABO and Rh Blood Groups	119
Activity 3	Hemostasis	123

Cardiovascular

Activity 1	Cardiac Cycle, Part 1	127
Activity 2	Cardiac Cycle, Part 2	135
Activity 3	Capillary Exchange	141
Activity 4	Hemodynamics	145

POGIL
WWW.POGIL.ORG
Copyright © 2015

Activity	Topic	Page

Immunity

Activity 1	Innate Immunity	153
Activity 2	Adaptive Immunity: T-cells and the Cellular Immune Response	159
Activity 3	Adaptive Immunity: The Humoral Response	165

Thoracic and Abdominal Viscera

Activity 1	Physiology of Upper GI Tract	171
Activity 2	Anatomy of Ventilation	179
Activity 3	Renin - Angiotensin System	185
Activity 4	Homeostasis	189
Activity 5	Acid/Base Homeostasis	191

Reproductive and Development

Activity 1	Meiosis	195
Activity 2	The Menstrual Cycle	201
Activity 3	Making a Person: From Zygote to Gastrula	205

Genetics

Activity 1	Mendelian Genetics	211
Activity 2	More Complex Forms of Inheritance	215

	Image Credits	**221**

The Language of Science and Medicine

"Why do scientists seem to speak a foreign language (in any language)?"

Model 1: Common Root Words and Their Meaning

Root word	Meaning	Example	Example Definition
arthr-	Joint	Arthritis	Inflammation of a joint
brachi-	Arm	Brachial	Having to do with the arm
card-	Heart	Endocarditis	Inflammation of the heart lining
cerv-	Neck	Cervical cancer	Cancer of the neck of the uterus
cyt-	Cell	Cytology	The study of cells
dactyl-	Fingers/toes	Polydactyly	Having too many fingers or toes
derm-	Skin	Dermatologist	Physician who specializes in the skin
gastr-	Stomach	Gastrin	Hormone secreted into stomach
hepat-	Liver	Hepatocyte	Liver cell
hydro-	Water	Hydrophobic	Water fearing
kal-	Potassium	Hypokalemia	Not enough potassium in the blood
my-, myo-	Muscle	Myalgia	Muscle pain
nephr-	Kidney	Nephropathy	Kidney disease
neur-	Nerve	Neuralgia	Nerve pain
onco-	Cancer	Oncologist	A physician who specializes in cancer
sept-	Contamination	Septicemia	Contamination in the blood
vas-	Vessel	Vasodilation	Enlargement of a blood vessel
natr-	Sodium	Natriuretic	Causing the excretion of sodium

Critical Thinking Questions

1.   What root word is used to indicate something that has to do with water?

2.   What roots could you use to describe a structure associated with muscle?

3.   Someone who has neuritis is having problems with what structure(s)?

4.   Vertebrae are the bones of the spine. Where are your cervical vertebrae?

POGIL
WWW.POGIL.ORG
Copyright © 2015

Application

5.   Based on the model above, your group should devise a short, grammatically correct, English sentence that defines the word cardiomyocyte. Be prepared to share your definition with the class.

6.   What suffix do arthritis and endocarditis have in common?

      a. Based on this observation, what does your group think the suffix – itis means?

      b. Likewise, what does –emia probably mean based on the examples in the model?

7.   Write brief descriptions (like in the model) of the following words:

      a. Hyponatremia:

      b. Dermatitis:

Model 2: Common Prefixes and Their Meaning

Prefix	Meaning	Example	Example Definition
a-	Without, not	Amenorrhea	Cessation of menstruation
dys-	Bad, wrong	Dysplasia	Problem with growth, malformation
ex-, exo-	Out of, out from	Exocytosis	To secrete from a cell
hem-, hemat-	Blood	Hemophilia	lit. "Love of bleeding", inability to clot
hyper-	Above, over	Hyperkalemia	Too much potassium in the blood
hypo-	Below, under	Hypothermia	Below optimal body temperature
para-	Near, next to	Paracrine	Secretions that target nearby tissues
quadri-	Four of	Quadriceps	Muscle with four parts

Critical Thinking Questions

8.   Based on the model, what does the prefix a- indicate?

9.   If a physician tells you that *Staphylococcus aureus* can cause an asymptomatic infection, what effect will that infection have on a patient?

10.   Based solely on the model, where could you expect to find your parathyroid glands?

Application

11.   As a group, estimate the body temperature of someone suffering from hyperthermia. Be able to justify your answer.

12.   The root word "glycol" means sugar. Based on this definition and the two prior models, write a definition below for the term hyperglycemia.

13.   You are a nurse in an obstetrics practice. A young couple has just been told that their child will be born suffering from adactyly. How would you explain this to them?

Model 3: Zeus

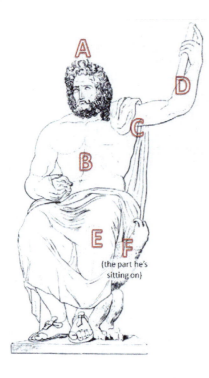

(the part he's sitting on)

Critical Thinking Questions

14. Use whatever words you like (above, in front of, top, etc) to describe the relative positions of the labeled areas indicated below:

a. A relative to B

b. C relative to B

c. D relative to C

d. E relative to B

e. F relative to B

When you have finished with Parts a-e, send your group's reporter to another group and compare answers. Make a note of how many answers are exactly the same.

15.   Now look at Model 4 on the next page. Using Model 4 as a guideline, answer Questions a-e again in the space below.

a. A relative to B

b. C relative to B

c. D relative to C

d. E relative to B

e. F relative to B

When you have finished with Parts a-e, send your group's reporter to another group and compare answers. Make a note of how many answers are exactly the same.

Is this more or less than the last time?

Model 4: Directions of the Body

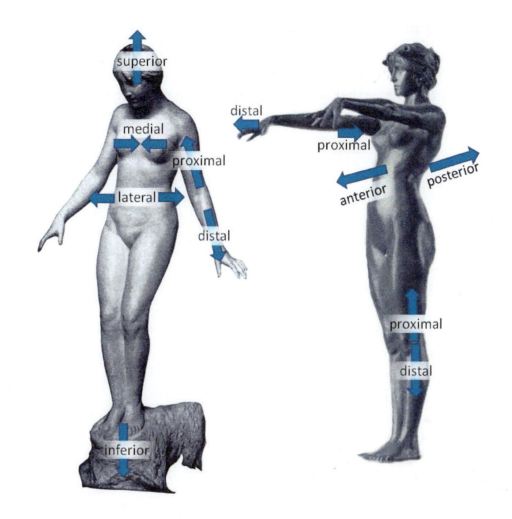

16.   We have learned quite a few specific medical/anatomical terms today. Take a few minutes and discuss in your group whether or not you see any value in having such specific terms in science and medicine. The scribe should summarize your discussion in the space below, the presenter should be able to defend your answers.

Exercises

Use the "http://www.globalrph.com/medterm.htm" to answer the following questions.

1.   Define these terms just using the words parts in your guide. Don't look them up in a glossary or online, that will totally defeat the purpose of this activity.

   a. adipsia

   b. hepatitis

   c. hypernatremia

   d. renomedullary

   e. histocytosis

   f. polyphagia

2.   Create a scientific term to describe the following:

   a. Creation of tissues

   b. White cell

   c. To cut the trachea

   d. Pain in the eyes

3.   Fill in the blanks below using the proper directional terms from Model 4 of this exercise.

   a. The elbow is _____ to the wrist.

   b. The breastbone is _____ to the spine.

   c. The nose is _____ to the cheeks.

   d. The ears are on the _____ aspect of the head.

   e. The ankle joint is_____ to the knee joint.

   f. The kneecap is on the_____ surface of the knee.

The Building Blocks of Matter

"What is the difference between an atom, ion, compound, and molecule?"

Model1: Some Atoms, Isotopes, and Ions of Various Elements

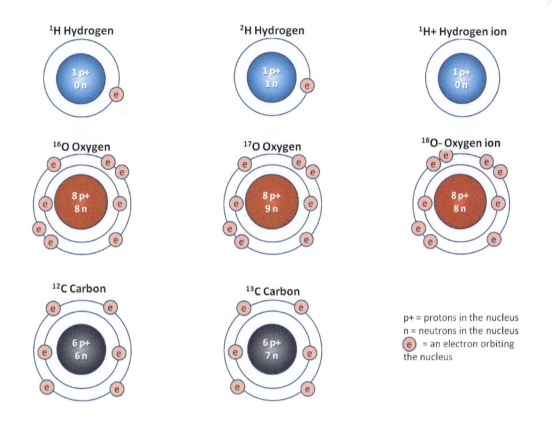

Critical Thinking Questions

1. According to the model, how many protons are in...

    a. An ^{1}H hydrogen atom, ^{2}H hydrogen atom, and ^{1}H+ hydrogen ion?

    b. An ^{16}O oxygen atom, ^{17}O oxygen atom, and ^{16}O- oxygen ion?

    c. A ^{12}C carbon atom and ^{13}C carbon atom?

2.   According to the model, how many neutrons are in...

    a. An ^1H hydrogen, ^2H hydrogen, and $^1H+$ hydrogen ion?

    b. An ^{16}O oxygen atom, ^{17}O oxygen atom, and $^{16}O-$ oxygen ion?

    c. A ^{12}C carbon atom and ^{13}C carbon atom?

3.   Based on your answers to Questions 1-2, what is the only thing that all hydrogens have in common?

    a. Do oxygen and carbon follow a similar pattern?

4.   Carbon is an element. Oxygen and hydrogen are also separate elements. <u>Based on your answers to Question 3</u>, write a complete sentence that defines an element.

5.   Based on the model, what is the difference between ^1H hydrogen atom and the H+ ion?

    a. What is the difference between ^{16}O and the O- ion?

6.   Based on your answers to Question 5, write a brief definition of an ion.

Application

7.   Why does the symbol for the hydrogen ion have a + sign and the symbol for the oxygen ion have a − sign?

Model 2: Molecules and Compounds

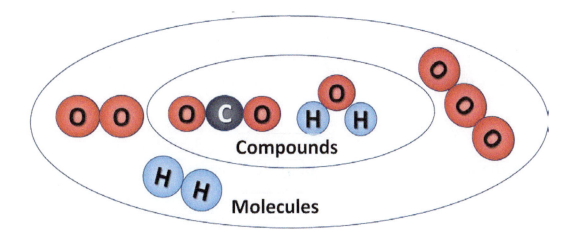

Critical Thinking Questions

8.   Everything in the large circle is a molecule. How many molecules are in Model 2?

        a. List the molecules from the model in the space below.

9.   How many atoms are in a molecule? Circle one of the following: exactly one, exactly two, exactly three, or more than one.

10.   How many compounds are in Model 2?

        a. List the compounds from the model in the space below.

11.   Are compounds always molecules, according to the model?

12.   Are molecules always compounds, according to the model?

13.   As a group, discuss the similarities and differences between atoms, molecules, and compounds. Respond to 'a' and 'b' below with an answer that reflects the group discussion.

a. In the space below, write a grammatically correct English sentence that defines a molecule.

a. In the space below, write a grammatically correct English sentence that defines a compound.

Application

14.   Using Models 1 and 2 as a guide, draw the following and label them as compounds and/or molecules (remember it is possible to be both).

a. Diatomic Nitrogen (N_2)

b. Nitrous Oxide (N2O)

c. Ammonia (NH3)

Exercises

1.   Go to http://www.ptable.com/ and use the interactive <u>Periodic Table of Elements</u> to locate the chemical symbol for the elements below and fill in the table.

Oxygen:	Carbon:	Nitrogen:	Calcium:	Phosphorus:	Potassium:
Sulfur:	Sodium:	Chlorine:	Magnesium:	Iron:	Iodine:

Memorize these symbols!

2.   Write the symbol for the compounds carbon dioxide and water in the space below:

Acids, Bases, and pH (Oh My!)

"What does pH mean?"

Model 1: The pH Scale

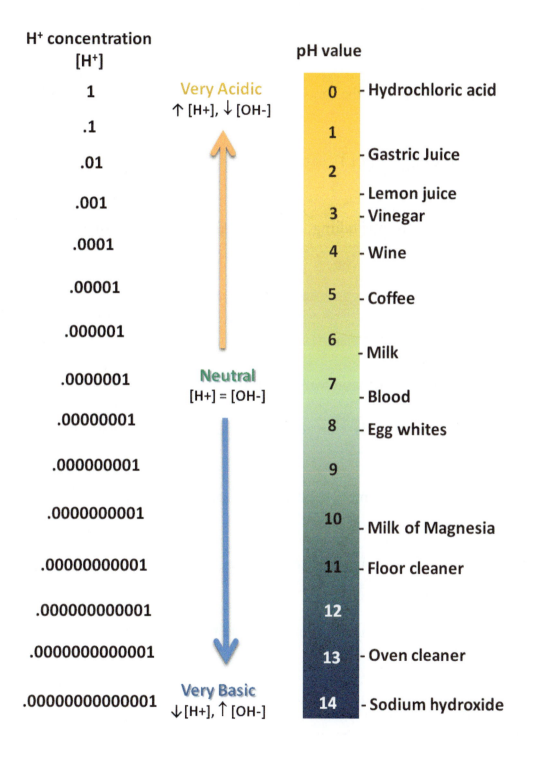

Critical Thinking Questions

1.   According to the model, do acids have low pH values or high pH values?

   a. What about bases, low values or high values?

2.   What is the neutral pH according to the model?

3.   When pH is low (e.g. 1), is the concentration of hydrogen ions [H+] low or high?

   a. When pH is high (e.g. 13), is the concentration of hydrogen ion [H+] low or high?

4.   If the pH of a solution is falling, is it becoming more acidic or more basic?

   a. Is the number of hydrogen ions going up or down?

5.   If the pH of a solution is rising, is it becoming more acidic or basic?

   a. Is the number of hydrogen ions going up or down?

6.   Which of the following is more acidic (circle the correct answer): lemon juice or egg whites?

7.   Which of the following is more basic (circle the correct answer): coffee or vinegar?

8.   Which of the following is more acidic (circle the correct answer): milk of magnesia or oven cleaner?

9.   Which of the following is more basic (circle the correct answer): pH 6 or pH 12

10.   Which of the following is more acidic (circle the correct answer): [H+] = 0.001 or 0.00001

11.   As a group, write a complete sentence that defines pH.

Application

12.   Human blood and tissue fluid is ALWAYS maintained at pH 7.4, no matter what. If hydrogen ions are removed from the blood, what might your body do to return the pH to 7.4?

Cell Transport Mechanisms

"How does stuff get into and out of a cell?"

Model 1: Diffusion

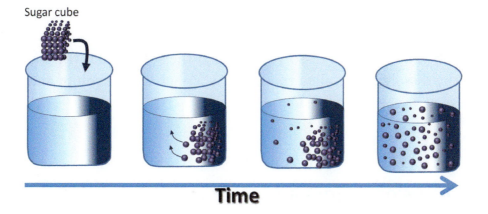

Sugar cube

Time

Critical Thinking Questions

1.   Circle the part of the model that represents time = 0 (the beginning).

     a. How are the sugar molecules (purple dots) arranged in this part of the model?

2.   In the next picture over, where are most of the sugar molecules located?

3.   In the third picture, circle the region of the beaker with the lowest concentration of sugar molecules. Draw a square around the area with the highest concentration of sugar molecules. Draw an arrow from the square to the circle.

     a. This beaker contains a **concentration gradient** of sugar indicated by your arrow. In the space below write a definition of 'concentration gradient' that everyone can agree on.

POGIL
WWW.POGIL.ORG
Copyright © 2015

4.   Look at the beaker that represents the time furthest from the beginning. How are the sugar molecules arranged in this part of the model? Does a concentration gradient still exist?

5.   If you put a packet of sugar into a glass of warm tea, stir it, and wait 5 minutes, what happens to the sugar?

    a. The substance that get dissolved is called the **solute**, and the substance doing the dissolving is called the **solvent**. What are the solute and solvent in the glass of sweet tea?

Model 2: Diffusion Across a Membrane

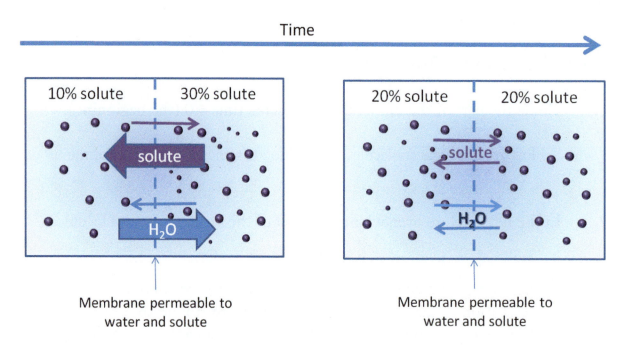

6.   According to Model 2, to what is the membrane permeable? [If you don't remember what permeable means, librarians can look it up].

7.   Does a concentration gradient exist in the figure on the left?

    a. If so, circle the portion of the figure that has a high concentration of solute.

8.   Compare Model 1 with Model 2. Are the end results (the right-most figure in both) essentially the same or are they different?

a. Would your answer be the same if the membrane were not permeable to the solute?

b. The process you have just explored is called **diffusion**. In the space below, write a group definition for **diffusion** that explains this process in simple, everyday language.

Model 3: What if We Can't Move Solute?

2 solute, 10 solvent   8 solute, 10 solvent

• Using 's' to represent solute and 'v' to represent solvent, fill in the boxes in the image on the left with the indicated numbers of 's' and 'v'
• Under each side of the box write the **concentration** of solute (solute divided by solvent)
• The dashed line represents a semi-permeable membrane. Solute cannot pass through, but solvent can. In the picture on the right, use 's' and 'v' again, but move 6 of the 'v's from the left side of the membrane to the right.
• Now write the concentration on each side of the membrane under the box on the right.

9.   The process we just observed in Model 3 is called **Osmosis**. Based on your answer to the questions above, perform the following tasks on Model 4.

a. On each of the drawings, indicate the direction water would have to flow in order for the concentration of solute (molecules of solute per molecules of solvent) to become equal. If no water flow is necessary, leave it blank.

b. Think about the prefixes hypo-, hyper-, and iso-. [Librarians can look these up if you need to refresh your memories]. Write the meaning of those prefixes in the space below.

c. **Tonicity** refers to the relative concentration of a solute on either side of a membrane. If the circles in Model 3 represent a cell, work with your group to determine if the interior of each cell is hypertonic, isotonic, or hypotonic. Label each cell <u>hypotonic, isotonic,</u> or <u>hypertonic</u> in the empty space above each cell in the model.

Model 4: Osmosis

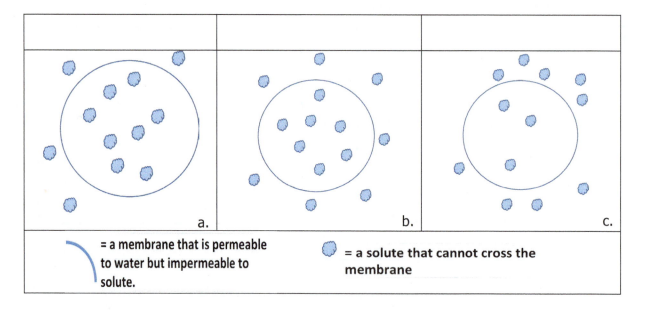

Model 5: What Crosses a Membrane

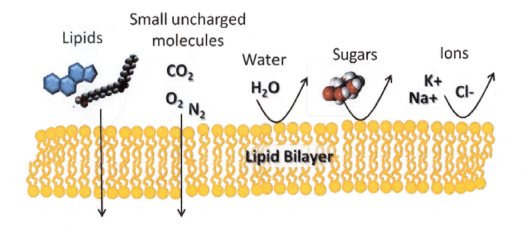

10.   Based on the model, what sorts of molecules can cross the lipid bilayer?

   a. What appears to get rejected by the membrane?

11.   Are lipids (fats and oils) hydrophobic or hydrophilic?

   a. Is the middle of the <u>lipid</u> bilayer hydrophobic or hydrophilic?

12.   What happens if you try to mix oil with water (think Italian salad dressing)?

13.   Based on your answers to Questions 10 – 12, devise a consensus answer to the following: Why can't hydrophilic molecules like sugars and ions pass through the lipid bilayer?

Model 6: Cellular Transport Processes

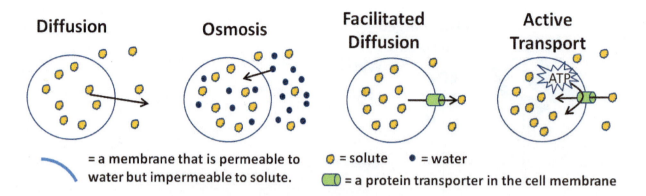

14. According to the model is solute concentration higher inside these cells or outside these cells?

15. Use Model 4 to fill in the table below.

	Diffusion	Osmosis	Facilitated Diffusion	Active Transport
What moves? (solute or water)				
Which way does it go? (from high solute concentration to low solute concentration or *vice versa*)				
Are membrane proteins required?				
Is energy in the form of ATP required?				

Exercises

1.   Match each of the following terms with their definitions.

_____ Concentration gradient	A. "Water loving" – substances that readily dissolve in water.
_____ Solute	B. "Water fearing" – substances that don't dissolve in water
_____ Solvent	C. When solute concentration is equal on both sides of a semi-permeable membrane.
_____ Diffusion	D. The movement of a solute across a membrane in the direction of its concentration gradient, but that requires the use of membrane proteins to traverse the membrane.
_____ Concentration	E. The amount of a solute per amount of solvent (e.g. grams/liter)
_____ Osmosis	F. A state in which there is an area of high solute concentration in one location and low solute concentration in another location
_____ Hypertonic	G. The movement of a solute across a membrane against its concentration gradient. This process requires both membrane proteins and energy.
_____ Hypotonic	H. Movement of a solute across a membrane in the direction of its concentration gradient (from high concentration to low concentration)
_____ Isotonic	I. When the concentration of solute is lower on the side of the membrane in question than the other.
_____ Hydrophobic	J. The substance that is dissolved.
_____ Hydrophilic	K. The movement of water across a membrane in order to equalize solute concentration.
_____ Facilitated Diffusion	L. The substance that dissolves other substances (e.g. water)
_____ Active Transport	M. When the concentration of solute is higher on the side of the membrane in question than the other.

Basic Histology

"What are the tissues that make up human organs?"

Model 1: There are 5 Tissues from Which All Organs are Constructed

Tissue Type	Basic Structure	Function	Examples
Epithelial Tissue 	Cells are tightly packed and there is little extracellular material. Can be found in sheets (surface epithelia) or glands.	Covers the surface of the body, lines body cavities, and provides protection. Absorption, secretion, and excretion.	
Connective Tissue	Most of the tissue is extracellular matrix (fibers and/or minerals), few cells.	Provides support to the body. Coordinates the physical activity between muscles and bones.	
Muscle Tissue	Cells are long and stringy, usually contain more than on nucleus, capable of shortening (contracting)	Movement... ..of the whole body ..of internal organs, blood, and some glands	
Nervous Tissue	Many different cells types with irregular shapes. Contains neurons (nerve cells) and neuroglia (supporting cells)	Communication between body parts. Coordination and regulation of body activities/functions.	
Blood	Contains red blood cells, white blood cells, and platelets in a liquid matrix (called plasma)	Transport of gases, nutrients, hormones, and wastes.	

Copyright © 2015

Critical Thinking Questions

1.   In the space below, list the **five basic tissues** found in the model.

2.   Compare epithelial tissue and connective tissue.

a. Which one has a higher density of cells?

b. Which one has more extracellular matrix?

3.   What organ best matches up with the first listed function of epithelial tissue?

4.   Librarians, use Wikipedia to look up each of these terms below and explain them to the rest of your group. As a group, discuss these and list at least one example of an organ that might meet each of the criteria and list it below.

Absorption:

Secretion:

Excretion:

5.   Based on the model, what kinds of things make up extracellular matrix (Hint: look at the structure of connective tissue)?

6.   Hydroxyapatite is a chemical that contains a great deal of calcium. Name a specific example of a structure in the human body that likely contains this **calcium-rich** mineral.

7.   In addition to the tissue you just listed, there is another connective tissue that plays a major role in support. Think about the tip of your nose or your ears and list that other tissue below.

8.   Look at the second function of connective tissue. The picture of connective tissue in the model fulfills this role. What structures in your body coordinate the actions of muscles with bone (like the thing that attaches your calf muscles to your heel)?

9.   Based on the model, what are the things that are moved by muscle tissue?

a. As a group, discuss places other than your musculoskeletal system where you might find muscle tissue and list at least two of those places in the space below.

10.   Looking at the function of nervous tissue, what organ do you think is responsible for coordinating and regulating all your body's activities and functions?

a. What does this organ use to communicate with the rest of the body (what are its communication cables)?

Application

11.   Based on your answers to Questions 1-10, fill in the fourth column of the model with examples of where you might find each tissue. You should have at least 4 examples for epithelial tissue, 3 for connective tissue, 3 for muscle tissue, 2 for nervous tissue, and 1 for blood. You may send your spokesperson to consult with other groups if you get stuck.

Exercises

Name the tissue that comprises each structure listed below:

1.   Ligaments:

2.   Membrane that covers the liver:

3.   Myometrium of the uterus (responsible for contractions during labor):

4.   Myenteric plexus (communicates between thoracic nerves and digestive tract:

5.   Intestinal submucosa (supports lining of GI tract):

6.   Nociceptor (detects pain and send signal to brain):

7.   Pyloric sphincter (contracts to close off stomach from intestines):

8.   Thyroid gland (secretes hormones):

9.   Alveolar cells (Absorb oxygen and excrete carbon dioxide):

10.   Intervertebral disks (support spine and cushion against compression):

Skin and Temperature Control

"How does skin help maintain a stable body temperature?"

Model 1: The Skin as Thermoregulator

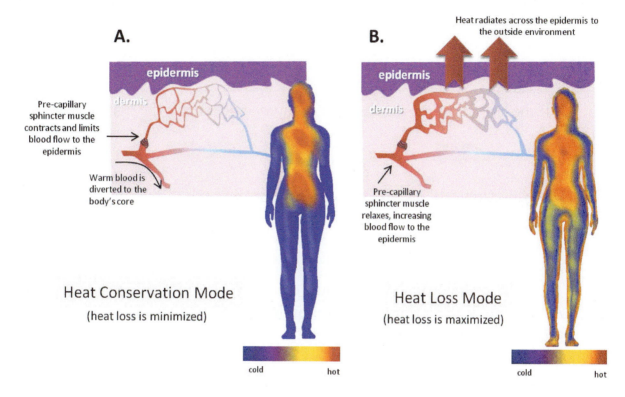

A.

Heat Conservation Mode
(heat loss is minimized)

B.

Heat radiates across the epidermis to the outside environment

Heat Loss Mode
(heat loss is maximized)

Critical Thinking Questions

1. In 'heat conservation mode' what have the precapillary sphincter muscles done?

    a. What has that action done to the amount of blood traveling to the epidermis?

    b. According to the model, this action causes blood flow to be diverted where?

2. If blood is kept deeper in the body, will the heat in the blood be able to escape easily?

    a. Why might that be a good idea in 'heat conservation' mode?

POGIL
WWW.POGIL.ORG
Copyright © 2015

3.   Look at Part 'B' of the model. What is different about blood flow in this part of the model compared to Part 'A'?

4.   What effect has this change in blood flow had on the distribution of heat in the body (look at the thermal image)?

a. If the left side of the model is 'heat conservation' what is the right side labeled?

b. When might we want to increase heat loss?

5.   The mechanism of heat loss you just discovered is called **radiation heat loss**. As a group devise <u>a simple description in everyday terms</u> that explains how the skin controls **radiation heat loss**.

6.   If you put an ice cube in warm water, will heat flow from the warm water into the ice cube and melt it, or will heat flow from the ice cube into the water and freeze it?

a. Based on your answer to the first part of Question 6, does heat move from an area of high heat to an area of low heat, or does heat flow from an area of low heat to an area of high heat?

b. If the air surrounding the skin is still (no wind) and higher than 37°C (98.6°F), how much heat is likely to be <u>radiated</u> away from the skin into the surrounding air (none, a little, or a lot)?

7.   If you were to walk outside and the temperature is higher than 99°F, do you automatically die?

a. If you were to walk outside into 99°F temperatures, what is the first thing you would likely notice on your skin (Particularly your head and underarms)?

b. What happens to liquids when they are exposed to heat (do they stay liquid)?

c. Besides increased blood flow to the skin, what other physiological process aids in heat loss? Why does this cool you down [Hint: what is the process called by which liquids become gases]?

8.   What happens to a spoon of hot soup when you blow on it?

a. Which feels cooler, a 90°F room with still air or with a fan on?

Convection is the movement of molecules within a fluid (blowing air or running water).

b. As a group, decide if air convection will likely increase or decrease the rate of evaporation from the skin?

9.   In the space below, list the three main ways that the body sheds excess heat.

Application

10.   Professor McSpeedy loves to run. Answer the following questions about heat generation and loss in runners.

a. What tissue type is going to be doing most of the work in <u>moving</u> Professor McSpeedy as he runs?

b. Based on your answer to 'a', what tissue is responsible for generating most the body heat generated during exercise?

c. When professor McSpeedy hits the three-mile mark he is really sweating. Besides evaporation, what are two other mechanisms by which his body is losing heat?

d. If Professor McSpeedy runs on a treadmill indoors, which of the three mechanisms will likely **not** play a factor in keeping him from overheating?

e. Is there some way to introduce this method while on a treadmill? How?

11.   If someone is suffering from heat stroke, would it be more effective to put them in a tub of cool water or put them in a cool shower? Explain your answer.

Model 2: Regulation of Body Temperature

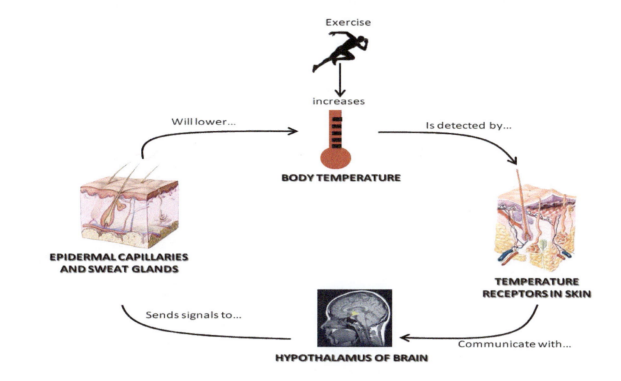

Critical Thinking Questions

12.   According to the model, what will cause an increase in body temperature?

a. What detects this change in temperature?

b. What is the body's response to this increase in temperature (what two actions will bring temperature down)?

c. What organ is **integrating** the information from the skin receptors and the body's normal set-point temperature?

13.   Body temperature is regulated by a **negative feedback loop**. There are four components to a negative feedback loop, a **variable**, **sensor**, **integrator**, and one or more effectors. Label the variable, sensor, integrator, and effectors on Model 2.

14.   The temperature in this room is regulated by a negative feedback loop.
In the space below, write the components of the room's heating/AC system
(thermometer, thermostat, room temperature, AC fan/heater) that
correspond to the variable, sensor, integrator, and effector.

Range of Motion

"How do you explain how a joint moves?"

Model 1: Movement of the Extremities in Anterior View

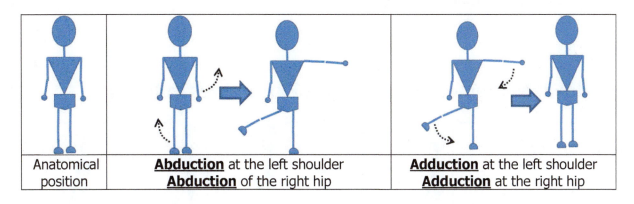

Anatomical position	**Abduction** at the left shoulder **Abduction** of the right hip	**Adduction** at the left shoulder **Adduction** at the right hip

Critical Thinking Questions

1. Based on the information in Model 1, when the left upper extremity is abducted, is it moving toward or away from the midline of the body?

2. When the right hip is <u>adducted</u>, is it moving towards the midline of the body or away from it?

3. Stand up and raise your left upper extremity to the side so that it is sticking straight out, parallel with the ground.

    Is that limb abducted or adducted? (Circle your answer)

4. Which movement would you have to make in order to return your left upper extremity to its starting position.

POGIL
WWW.POGIL.ORG
Copyright © 2015

Application

5.   Someone in your group do the following. Start in a standing position with your arms at your side, then simultaneously **abduct** all four extremities at once, then immediately return to starting position. What exercise have you just performed?

Model 2: Movement of the <u>Right</u> Extremities in Profile

Right shoulder (arm)	Extension	Flexion	Extension	Extension	Extension
Right elbow (forearm)	Extension	Extension	Flexion	Extension	Extension
Right hip (thigh)	Extension	Extension	Extension	Flexion	Extension
Right knee (leg)	Extension	Extension	Extension	Extension	Flexion

Critical Thinking Questions

6.   Circle the joint on each picture that is flexed (if no joints are flexed, leave it blank).

7.   Using the information in Model 2, fill in the table below.

Right shoulder (arm)			
Right elbow (forearm)			
Right hip (thigh)			
Right knee (leg)			

8.   As a group, fill in the blank with the function of the following muscles (then perform that action yourself):

Flexor digitorum muscle – _____ the digits

Adductor magnus muscle – _____ the thigh

Extensor indices muscle - _____ the index finger

Model 3: Special Movements of the Hands and Feet

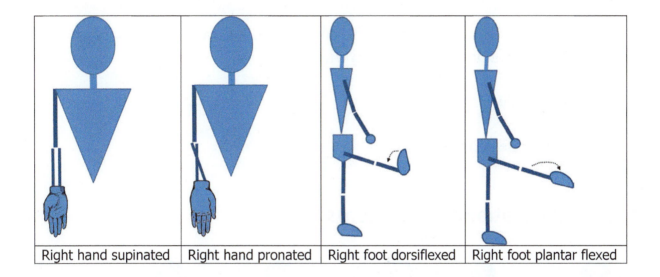

| Right hand supinated | Right hand pronated | Right foot dorsiflexed | Right foot plantar flexed |

9.   When one is standing in anatomical position are the hands pronated or supinated?

10.   What hand motion do we most commonly associate with looking at a wristwatch?

Ballet	Contemporary

11.   Look at the joints indicated in the table above. According to the table, which style of dance is associated with…

   …dorsiflexion?

   …plantar flexion?

Model 4: Rotational Movements

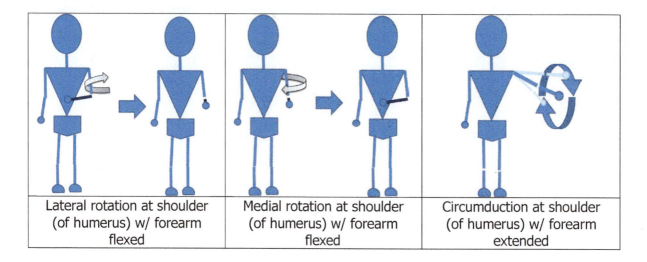

Lateral rotation at shoulder (of humerus) w/ forearm flexed	Medial rotation at shoulder (of humerus) w/ forearm flexed	Circumduction at shoulder (of humerus) w/ forearm extended

12.   According to the model, in which direction does medial rotation turn the extremity, towards or away from the midline?

13.   According to the model, in which direction does lateral rotation turn the extremity, towards or away from the midline?

14.   As a group determine what medial and lateral rotation of the thigh (at the hip) would look like. Be prepared to demonstrate for the class.

15.   Which of the following geometrical figures most closely resembles the path made by a joint undergoing circumduction? Circle your answer.

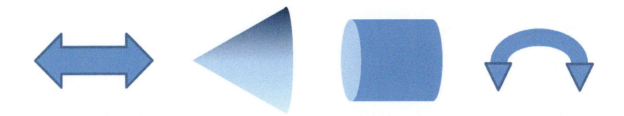

16.   As a group determine what circumduction of the thigh (at the hip) would look like. Be prepared to demonstrate for the class.

Application (we dance!)

17.   Follow the instructions below. Work as a group to figure out these sweet dance moves!*

a. Begin standing with your upper extremities at your side and your feet shoulder width a part.

b. Pronate your right hand as you flex your right arm to 90°.

c. Pronate your left hand as you flex your left arm to 90°.

d. Supinate your right hand.

e. Supinate your left hand.

f. Flex your right forearm <u>at the same time</u> as you medially rotate the right arm until your right hand rests on the opposite shoulder.

g. Flex your left forearm <u>at the same time</u> as you medially rotate the left arm until your left hand rests on the opposite shoulder.

h. With your elbows remaining in flexion, flex further at the left shoulder until your left hand is behind your head.

i. With your elbows remaining in flexion, flex further at the right shoulder until your right hand is behind your head.

j. With your elbows remaining in flexion, extend at the left shoulder until your left hand rests on your right hip.

k. With your elbows remaining in flexion, extend at the right shoulder until your right hand rests on your left hip.

l. Adduct the right arm as you supinate the right hand.

m. Adduct the left arm as you supinate the left hand.

n. Circumduct both hips simultaneously three times.

Ossification

"How is bone made?"

Model 1: Intramembranous Ossification

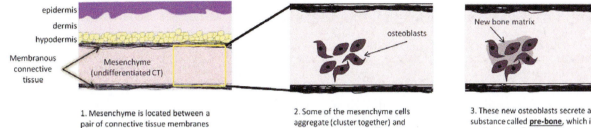

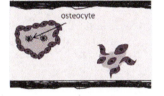

1. Mesenchyme is located between a pair of connective tissue membranes

2. Some of the mesenchyme cells aggregate (cluster together) and differentiate into **osteoblasts**

3. These new osteoblasts secrete a substance called **pre-bone**, which is made mostly of collagen fibers. Pre-bone is then calcified into bone matrix.

4. The resulting "island" of bone is called an **ossification center**. Osteoblasts become **osteocytes** when they are surrounded by bone matrix. Additional ossification centers form elsewhere in the membrane.

5. Multiple ossification centers grow toward one another and merge together to form **spongy bone**.

6. In the fully mature bone the spongy bone (or **diploë**) separates an **inner and outer table** comprised of compact bone.

Critical Thinking Questions

1. According to the model, what tissue is also called undifferentiated connective tissue?

2. What action do the cells of this tissue take immediately before they differentiate into osteoblasts?

3. What substance do osteoblasts secrete?

   a. What is that substance made of?

   b. What has to happen to that substance before it becomes mature bone matrix?

4.   What is the term for these clusters of osteoblasts that create "islands" of bone?

        a. Does it appear that the entire bone grows from a single one of these clusters?

        b. What, then, forms the mature flat bone?

5.   Looking at the picture of the mature bone, where is the only place you still see osteoblasts (the small oblong cells)?

        a. Based on your previous answer, write a group explanation as to how the bone might continue to grow once it's shape has been established.

Application

6.   Cleidocranial dysplasia is a disorder in which some mesenchyme cells are prevented from transforming directly into osteoblasts, resulting in a decrease in osteoblastic activity in developing flat bones. In the space below, list some of the bones that might be affected by this disorder.

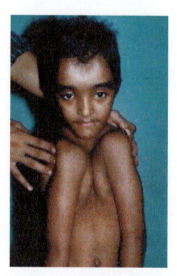

        a. What bones are likely deformed or missing in this young man suffering from cleidocranial dysplasia? [You can use the common name if you want]

Image courtesy of St. Joseph's Pediatric Dentistry Lit. Review

Model 2: Endochondral Ossification, Part 1

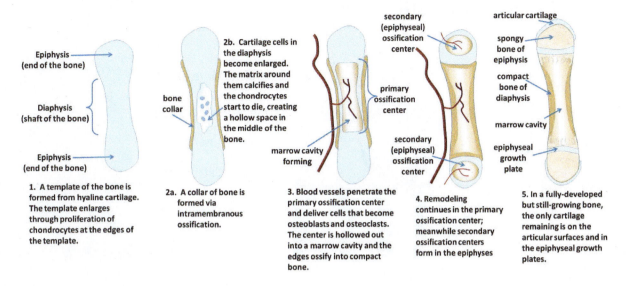

1. A template of the bone is formed from hyaline cartilage. The template enlarges through proliferation of chondrocytes at the edges of the template.

2a. A collar of bone is formed via intramembranous ossification.

2b. Cartilage cells in the diaphysis become enlarged. The matrix around them calcifies and the chondrocytes start to die, creating a hollow space in the middle of the bone.

3. Blood vessels penetrate the primary ossification center and deliver cells that become osteoblasts and osteoclasts. The center is hollowed out into a marrow cavity and the edges ossify into compact bone.

4. Remodeling continues in the primary ossification center; meanwhile secondary ossification centers form in the epiphyses

5. In a fully-developed but still-growing bone, the only cartilage remaining is on the articular surfaces and in the epiphyseal growth plates.

Critical Thinking Questions

7.   According to the Model 2, what forms the template or model for a developing long bone?

8.   What tissue is laid down as a collar around this model?

9.   Based on what you learned from Model 1, what do you think is going to happen to the cartilage in the primary ossification center?

10.   In what portion of this *long bone* is the secondary ossification center located?

a. What is the name of the cartilaginous structure that separates the epiphysis from the diaphysis?

11.   Look back at Model 1, does it appear that flat bones begin with a cartilage template?

12.   In the space below, compose a sentence as a group that best describes the difference between intramembranous and endochondral ossification.

Model 3: Endochondral Ossification, Part 2
The Epiphyseal Growth Plate

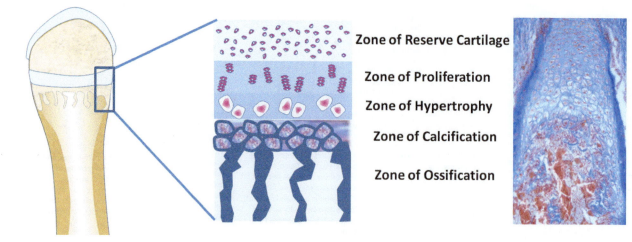

13.   Chondrocytes are cartilage cells. If these cells are proliferating, what does that mean they are doing?

14.   Based on the order of the various zones in the growth plate, what does it appear happens to the cartilage cells (chondrocytes) after they proliferate?

15.   Hyper- means more than normal, and trophy- means to eat. What happens to a person if they eat more than necessary for an extended period of time?

       a. Based on your previous answer and the appearance of the chondrocytes in the zone of hypertrophy, what does hypertrophy mean?

16.   What happens to the cartilage after it hypertrophies?

17.   What does the calcified cartilage become in the ossification zone?

Application

18.   Draw an arrow next to the growth plate in Model 3 to indicate the direction in which the bone is growing.

Exercise

1.   On a clean sheet of paper draw two side-by-side flow charts the show the process of intramembranous and endochondral ossification. It is OK to use your textbook for this.

Articulations

"How does the anatomy of a joint relate to the way that joint moves?"

Model 1: Primary Joint Classifications

Functional Name	Structural Name	Degree of Movement Permitted
Synarthroses	Fibrous	Immovable
Amphiarthorises	Cartilaginous	Only slightly movable
Diarthrosis	Synovial	Freely movable

Critical Thinking Questions

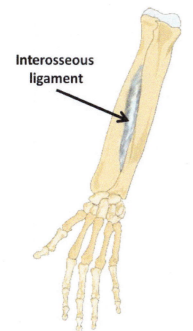

Interosseous ligament

1. What are the two ways that an immovable joint can be classified?

2. What tissue does your group think forms the juncture between bones in an amphiarthrosis?

3. What is the range of movement permitted by a synovial joint?

4. What connective tissue is joining the radius and ulna to the left?

5.   How would you classify this joint structurally?

6.   Is this a synarthrosis, amphiarthrosis, or diarthrosis?

7.   What connective tissue is joining the ribs to the sternum in the picture to the right?

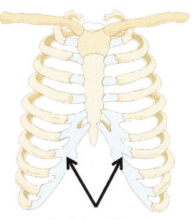

Costal cartilage

POGIL
WWW.POGIL.ORG
Copyright © 2015

8.   How would you classify this joint structurally?

9.   Is this a synarthrosis, amphiarthrosis, or diarthrosis?

10.   Stand up and move around a little. Then as a group, come up with at least 5 joints that are freely moveable (they allow the bones they connect to move across a wide range of motion in one or more planes), and list them in the space below.

Model 2: Classification of Synovial Joints

Types	Examples	Structure	Movement
Uniaxial			*Along one axis*
Hinge		Barrel-shaped process fits inside a trough-like cup.	
Pivot		Peg in a hole.	
Biaxial			*Along two axes*
Saddle		Two saddle-shaped turned 90° to each other.	
Condyloid		Oval shaped extension (condyle) fits into elliptical socket.	
Mulitaxial			*Around many axes*
Ball and Socket		Self-explanatory.	
Gliding		Two flat surfaces rubbing together.	

11.   Flex your arm at the elbow (bring your wrist toward your shoulder).

a. How many axes is your elbow able to move through?

b. Write 'elbow' into the appropriate box in the example column of Model 2.

12.   Look to your left. Now look to your right. Now to the left again. What kind of joint is your head sitting on? Write it in the appropriate box in the example column of Model 2.

13.   Condyloid joints can move quite far in one plane, but only a small way in the opposite plane. Flex and extend your wrist (hold out your arm in front of you and point your fingers to the ceiling and then to the floor). Then point your arm straight out and move your hand right to left at the wrist.

a. Does your wrist move as far side-to-side as it did up and down?

b. Does this suggest that the wrist joint is condyloid or ball and socket?

c. Write the word 'wrist' in the appropriate box in the example column of Model 2.

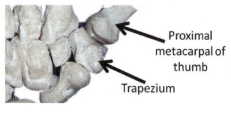

14. Look at this picture of the joint between the 1st metacarpal of the thumb and the trapezium bone. Does the trapezium resemble something you'd put on a horse? Write the word 'thumb' in the appropriate box in the example column of Model 2.

15.   Move your shoulder around. Based on its mobility, place 'shoulder' in the appropriate box in the example column of Model 2.

16.   The bones of your ankle are small and flat. Move your ankle through its entire range of motion. Based on the nature of the bones, the mobility of your ankle, and the fact that there is only one box left, write the word 'ankle' in the appropriate box in the example column of Model 2.

Application

18. For each of the joints listed below, classify it as a synarthrosis, amphiarthrosis, or diarthrosis. If it is a diarthrosis, further classify it as a hinge, pivot, saddle, condyloid, ball and socket, or gliding joint.

Hip	Knee	Cranial sutures	Pubic symphysis	Phalangeal (toes)
Vertebral (between the bodies)	Vertebral (between the articular processes)	Jaw	Tibio-fibular	Tooth-jawbone

The Sliding Filament Theory
"How do muscle cells contract?"

Model 1: Muscle Contraction Takes Place at the Level of the Sarcomere

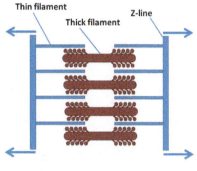

Relaxed Sarcomere   Contracted Sarcomere

Critical Thinking Questions

1. What structural component of the sarcomere is associated with arrows in Model 1?

2. Based on this, which component of the sarcomere is actually being moved when the sarcomere contracts and relaxes?

3. Which component of the sarcomere is physically attached to the structure that gets moved (i.e. the answer to Question 2)?

4. What component of the sarcomere is NOT directly attached to the Z-line?

5. Based on your answer to Question 4, what component of the sarcomere is likely pulling on the thin filaments to bring the Z-lines closer together?

POGIL
WWW.POGIL.ORG
Copyright © 2015

Application

6.   In the space below, write a short description as a group that explains the role of the thin filaments, thick filaments, and Z-line in sarcomere contraction.

Thin filament:

Thick Filament:

Z-line:

Model 2: Molecular Events of the Contraction Cycle

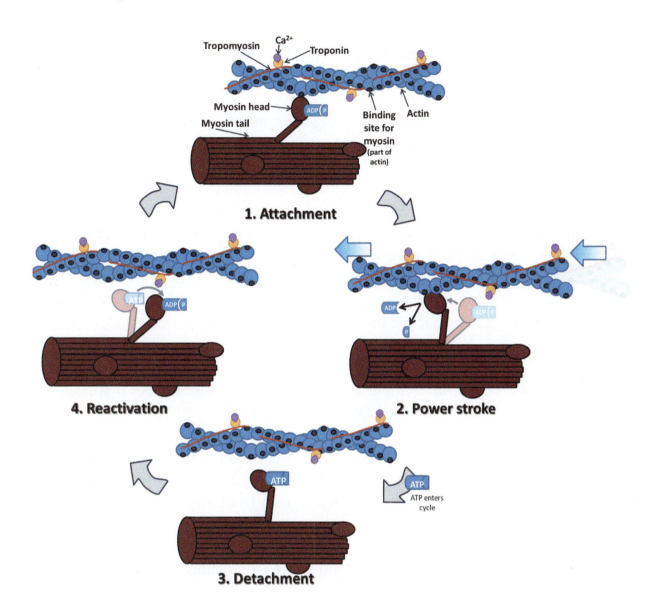

Critical Thinking Questions

7.   Find the thin filament in Part 1 of Model 2 (attachment) and draw a bracket and label it 'thin filament'.

8.   Which component of the thin filament makes the main 'string-of-pearls' portion of the filament?

       a. in the space below, list the other molecules that are found in/on the thin filament.

9.   The myosin head is bound to some other molecules, what are they?

      a. To which specific region of actin is the myosin head bound?

10.   In your own words, describe the change that occurs in the myosin molecule between Stage 1 (attachment) and Stage 2 (power stroke).

11.   The myosin head doesn't release from the actin binding site until a new molecule enters the cycle. According to the model, what molecule allows myosin to release from actin?

12.   In stage 4 of this cycle, the myosin molecule moves back into the **cocked position**. Where does it get the energy to re-cock itself?

13.   Look at all the myosin binding sites on the actin filament. Does it appear that each site lines up perfectly with a myosin head?

      a. Myosin can only bind to a binding site that is at the tip of a helix in the actin filament (like the one in Phase 1 of the model). After the power stroke (pulling Phase) and detachment is the myosin head lined up with a binding site at the tip of the filament?

      b. Can this particular myosin head bind to the actin filament for the next stroke?

      c. Look at the feet of these children. Are they all able to pull at exactly the same time?

      d. As a group, explain how myosin pulling on actin works kind of like kids playing tug-of-war.

Model 3: Effects of Calcium on the Thin Filament

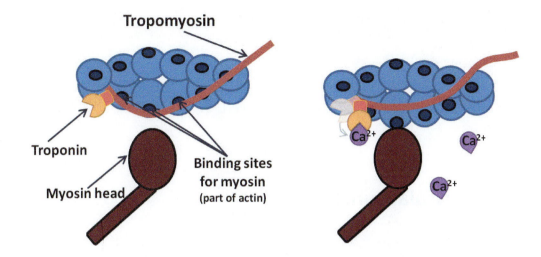

14.   Look at the left half of the model. Is calcium present here?

       a. Is the myosin binding site of actin exposed so that myosin can actually bind to it?

       b. What is blocking myosin's access to the binding site on actin?

15.   Now look at the right side of the model. Is calcium present here?

       a. To what molecule did it bind?

       b. What changed about the troponin molecule after it bound calcium?

       c. What effect did this change in troponin have on tropomyosin?

       d. Is the myosin binding site of actin exposed when calcium is present?

16.   In the space below, write a grammatically correct sentence (in English) that explains the role of troponin, tropomyosin, and calcium in muscle cell contraction.

Application

17.   What would happen to a muscle cell in either of the following situations?

a. The cell runs out of ATP.

b. There is an excess amount of calcium in the cell.

Model 4: Neuromuscular Junction

18.   Can muscles contract if there isn't any free calcium in the cell?

a. According to Model 5, where is calcium stored inside a muscle cell?

b. What causes the calcium to be released from where it is stored?

19.   What chemical messenger transmits the electrical impulse between the nerve cell and the muscle cell?

Application

20.   Acetylcholine esterase is an enzyme that removes Ach from the neuromuscular junction. The nerve gas Sarin blocks this enzyme causing Ach to remain permanently bound to the muscle cell receptors.

a. If Ach stays bound to its receptor, how will that affect Ca^{2+} levels inside the cell?

b. Would the cell be able to moderate whether or not it contracts or relaxes (i.e. would it be stuck in either the contracted or relaxed state)?

21.   As a group, explain why Sarin gas is a deadly poison.
[Hint: your diaphragm is a muscle]

Exercises

1.   Use Models 1-4 to put the following events in order from the signal from the brain reaching a muscle to the contraction of the whole muscle.

____   Myosin bends in two places, releasing ADP and pulling on the thin filament

____   The nerve impulse reaches the end of the nerve and causes it to release acetylcholine (Ach)

____   An electrical impulse travels down a nerve fiber

____   The Z-lines are pulled closer together and the A-band shrinks

____   Ach binds to receptors on the muscle cell membrane and causes the electrical impulse to be transmitted to the muscle cell

____   Calcium ions bind to troponin causing it to rotate

____   The myofibril gets shorter (contracts)

____   The electrical impulse inside the muscle cell causes the release of calcium ions from the endoplasmic reticulum

____   Rotation of troponin move tropomyosin off of the myosin binding site on actin

____   The myosin head binds the myosin binding domain of actin

Membrane Potentials

"Why are some cells electrically active?"

Model 1: The Sodium/Potassium Pump

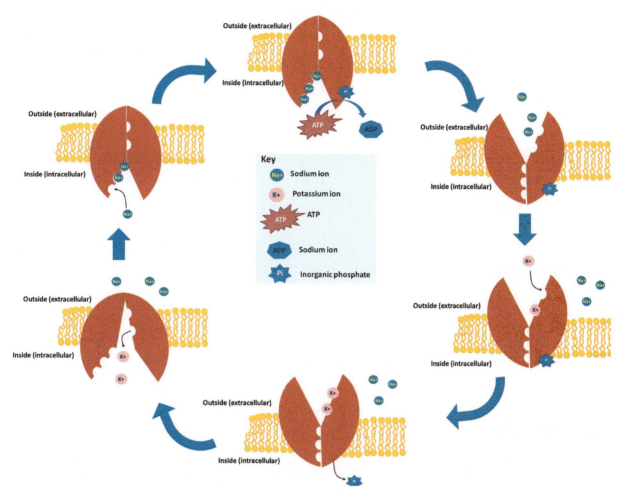

Key

- Na+ — Sodium ion
- K+ — Potassium ion
- ATP — ATP
- ADP — Sodium ion
- Pi — Inorganic phosphate

Critical Thinking Questions

1. What ion is being moved <u>out</u> of the cell according to Model 1?

    a. How many of these are being moved out?

    b. Is this ion positively or negatively charged?

2. What ion is being brought <u>into</u> the cell according to Model 1?

    a. How many of these are being moved in?

    b. Is this ion positively or negatively charged?

POGIL
WWW.POGIL.ORG
Copyright © 2015

3.   After this pump runs 5 times, how many positive ions will be moved outside the cell?

a. How many positive ions will have been brought into the cell after 5 cycles of the pump?

b. What is the difference in positive ions between the inside and the outside of the cell after 5 cycles?

c. What is the difference in positive ions between the inside and the outside of the cell after 70 cycles?

4. Imagine instead of pumping ions, we are pumping dollars. Is the cell spending more (pumping out) than it brings in?

a. If you spend more than you bring in, what happens to your bank balance?

Application

5.   In the space below draw a line to represent the cell membrane. Label one side 'inside' and the other side 'outside'. Place + symbols on the side of the membrane that would have the positive bank account (according to our analogy above) and put − symbols on the side of the membrane that would have the negative bank balance.

Model 2: A Polarized Membrane

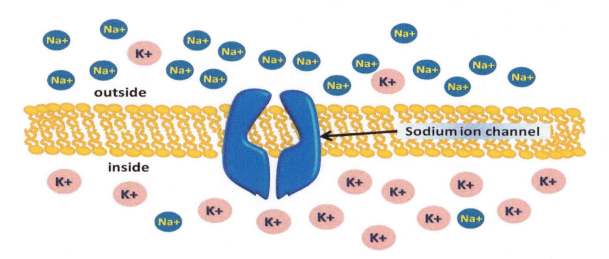

Critical Thinking Questions

6.   Place a large + sign on the side of the membrane with the most positive charges and a − sign on the side of the membrane with the least number of positive charges.

7.   Where are most of the sodium ions, inside or outside the cell?

8.   If the channel were to open, which way would sodium travel through the channel based on the laws of diffusion?

9.   What effect would this have on the **membrane potential** (the difference in charge across the membrane)?

Application

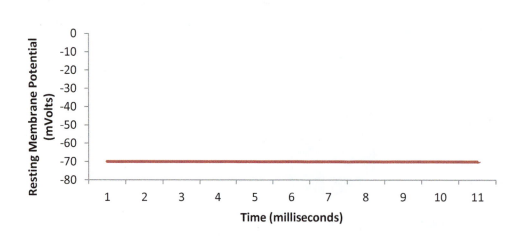

10.   This chart shows the **resting membrane potential** of a nerve cell. Notice that the line is running at around -70 millivolts.

a. The resting membrane potential has a negative value. Based on what we've already learned - is this measuring the inside or outside of the cell? [i.e. Which one is negative with respect to the other?]

b. If a sodium channel opens, will the inside of the cell become more positive or more negative as the sodium ions come in along their concentration gradient?

c. Based on your answer to 'a' and 'b', extend the line in the chart to show what might happen to the potential when sodium channels open.

d. Before the sodium channel opened, the membrane was polarized, like a battery (a positive and negative side). After opening the sodium channel, is the cell going to be polarized anymore?

Model 3: A Depolarized Membrane

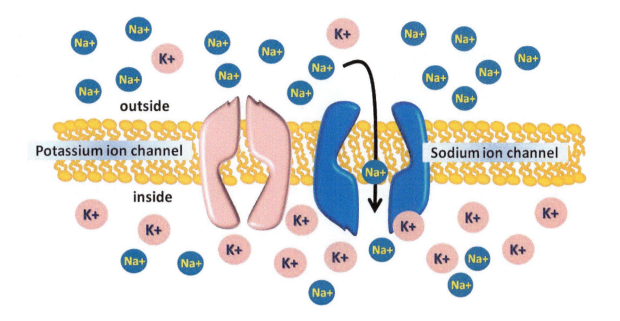

Critical Thinking Questions

11. What has changed about the sodium channel in this model (compared to model)?

12. How does the number of positive ions inside the cell compare with the number outside the cell?

13. Compare this model to Model 2 which contained a **polarized** membrane. Based on the ion distribution in these two models, devise definitions as a group, for the following:

Polarized

Depolarized

Application

14.   Based on the concentration of potassium ions inside and outside the cell, which way do you think K+ ions would go if the potassium ion channel were to open and the sodium channel were to close? [Hint: draw it if you need to]

15.   Would this depolarize or **repolarize** the membrane? Explain your answer.

Exercises

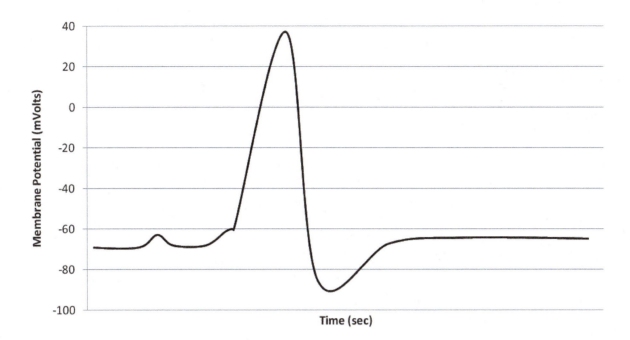

1. Find the points on this graph where membrane potential went up but went back down again without causing a drastic change in the polarity of the cell. Label that portion of the graph **Local Depolarization**.

2. Draw an arrow at the place on the graph where the entire membrane started to depolarize [Hint: it's when potential starts to rise as fast as it can].

   a. About what voltage was achieved before the membrane started to actually depolarize (i.e. where your arrow is pointing)

   b. **Threshold** is also called 'the point of no return', and refers to the voltage at which the entire membrane will depolarize. Based on your answers to 1 – 2a, draw a dotted line across the membrane at the <u>voltage</u> that represents the <u>threshold voltage</u>.

3. Circle the point on the graph where the potassium channels open and the sodium channels close.

4. Sometimes the potassium channels stay open long enough that the membrane will **hyperpolarize**. Indicate on the graph where the membrane is hyperpolarize. Indicate on the graph where the membrane is hyperpolarized.

Conduction of Action Potentials and Synapses

"How does an action potential get from one cell to another?"

Model 1: There are multiple types of ion channels in excitable cells.

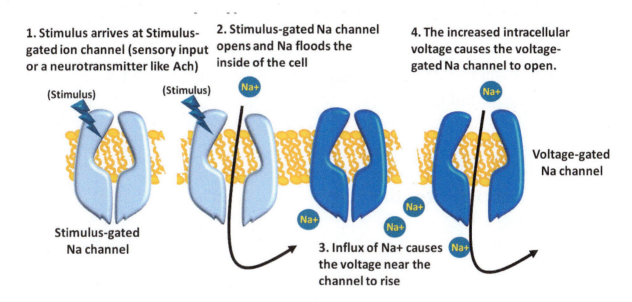

1. Stimulus arrives at Stimulus-gated ion channel (sensory input or a neurotransmitter like Ach)

2. Stimulus-gated Na channel opens and Na floods the inside of the cell

4. The increased intracellular voltage causes the voltage-gated Na channel to open.

(Stimulus)

(Stimulus)

Na+

Na+

Voltage-gated Na channel

Stimulus-gated Na channel

Na+

Na+

Na+

3. Influx of Na+ causes the voltage near the channel to rise

Critical Thinking Questions

1.   According to the model, what are some examples of stimuli that can cause a stimulus-gated Na channel to open?

2.   According to the model, what causes the voltage-gated Na channels to open?

      a. Small increases in potential will not cause these channels to open. What do you (the group) think is the term for the actual potential value that will cause these channels to open (the point of no return)?

Application

3.   Imagine a whole string of voltage-gated sodium channels lined up in a row on a membrane. As the first one reaches threshold and opens, what will happen to the voltage of the cell near that channel?

a. What will that do to the voltage-gated Na channel next to it?

b. What will that do to the voltage-gated Na channel next to that one? (Do you see a pattern here?)

Model 2: Action Potential

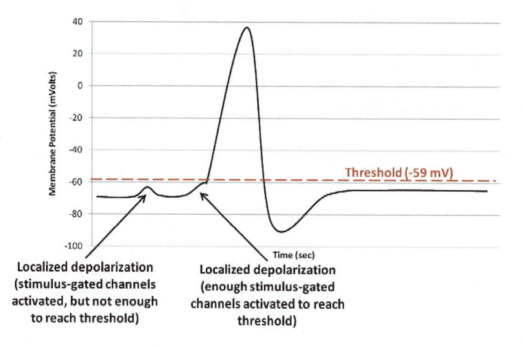

Critical Thinking Questions

4.   According to the model, what causes localized depolarization that doesn't result in the cell reaching threshold?

a. Do you think that the nerve cell is actively conducting a signal when these events occur?

5.   Indicate on the graph where the **voltage-gated sodium channels** come open.

6.   There are also **voltage-gated potassium channels** on these membranes. Indicate where on the graph these channels come open (Hint: think about what **repolarizes** the cell).

   a. Indicate on the graph where these voltage-gated potassium channels close. They should indicate minimal (hyperpolarized) voltage

Application

7.   A nerve cell is like a light switch, it's either actively conducting or it's not. Using the picture to the right as a stand-in for a neuron, show which positions indicate that the neuron is at

   1. Resting membrane potential

   2. Localized depolarizations

   3. Action potential

Model 3: Action Potential Propagation

Critical Thinking Questions

8.   What portion of the neuron is being highlighted in this model?

9.   Draw an arrow on the small picture of the neuron to indicate the direction the action potential is traveling.

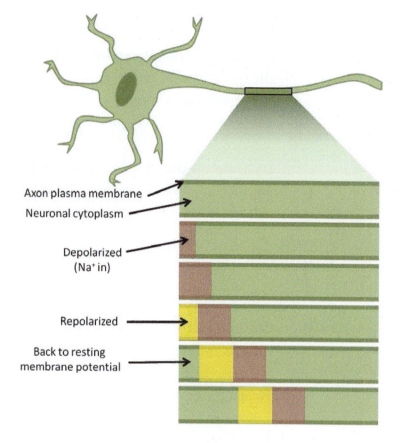

Axon plasma membrane
Neuronal cytoplasm

Depolarized (Na⁺ in)

Repolarized

Back to resting membrane potential

Application

10.   Based on this model and the first model, write a consensus explanation for the mechanism by which an action potential travels from the cell body (soma) down the axon.

Model 4: A Synapse

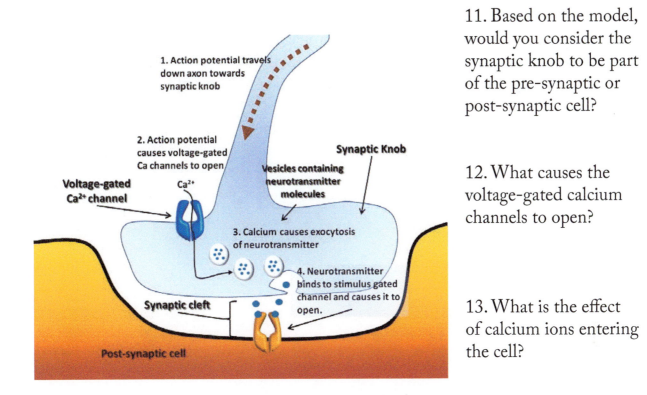

11. Based on the model, would you consider the synaptic knob to be part of the pre-synaptic or post-synaptic cell?

12. What causes the voltage-gated calcium channels to open?

13. What is the effect of calcium ions entering the cell?

14.   What kind of channel opens in response to binding a neurotransmitter?

Application

15.   If the channel in the post-synaptic cell were a K+ channel, would opening it stimulate or inhibit the post-synaptic cell?

Exercise

1.   On a clean sheet of paper, create a flow chart that illustrates all the events involved in a cell reaching threshold, propagation of the action potential down the axon of the cell, and transmission of the signal to the next cell in the pathway (the post-synaptic cell). You may use your textbook for this.

Everyone should work individually and we will compare your flowcharts together.

Reflex Arcs – The Simplest Neural Circuits

"Why do I pull my hand back from a hot stove before I even feel the heat?"

Model 1: A Somatic Reflex Arc

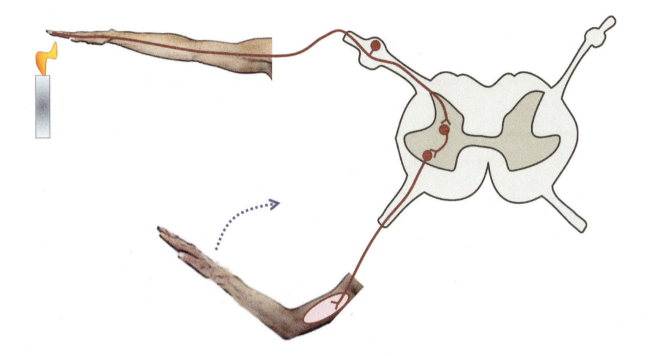

Critical Thinking Questions

1. Reflexes are an automatic **response** to a **stimulus**. According to the model, what is the stimulus?

    a. What is the response?

2. Does it appear, based on the model, that the brain is involved at all in this particular reflex action?

POGIL
WWW.POGIL.ORG
Copyright © 2015

3.   Somatic reflex arcs have five essential components. Based on your knowledge of vocabulary label the following five components in Model 1:

- Receptor
- Sensory neuron
- Integrating center
- Motor neuron
- Effector

Model 2: A Visceral Autonomic Reflex Arc

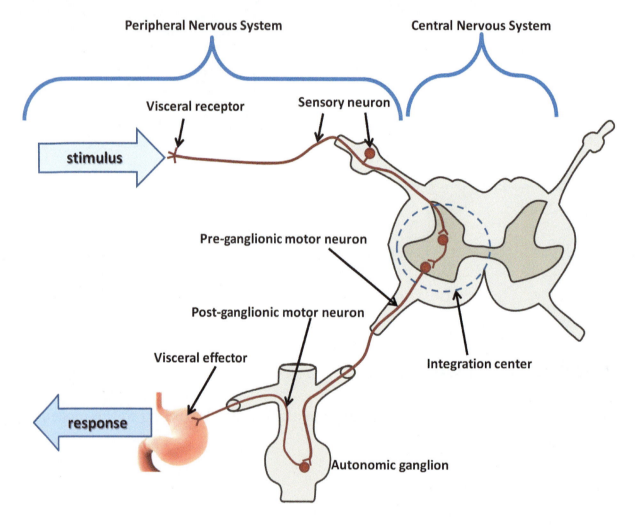

4.   Based on this model, what are the primary structural differences between a somatic and visceral reflex arc (what does this one have that the first one didn't)?

5.   According to this model, where is the integration center located? [Be as specific as possible]

6. Is it possible to have some visceral reflex arcs with integration centers in the brain?

7. In the space below, list at least three locations in the brain that are likely to serve as integration centers for <u>visceral</u> reflex arcs (i.e. they receive inputs and communicate out to the internal organs).

Application

8. Some people suffer from a condition known as congenital insensitivity to pain, in which they cannot <u>detect</u> painful stimuli at all, yet all their motor skills are normal. Which portion(s) of the somatic reflex arc might be defective in people with this condition?

9. There is a related condition known as congenital indifference to pain in which people can detect painful stimuli but they don't react properly to that stimulus. Essentially, these people don't feel pain as all that painful. People with this condition, like people with congenital insensitivity to pain, have normal motor function. Which portion of the somatic reflex arc might be defective in these people?

10. When your small intestine begins to fill with food, a signal is sent to the brainstem which initiates contraction of the smooth muscle in the abdominal wall and movement of the food through the intestines.

    a. Is this an example of a somatic or visceral reflex arc?

Exercises

1.   In the space below, fill in the table with the appropriate components of somatic and visceral reflex arcs.

	Visceral Reflex	Somatic Reflex
Input components		
Processing Component		
Output components		

Receptors, Receptors, Receptors

"How does our body perceive its surroundings?"

Model 1: Classification by Location

There are three types of receptors based on where they are located in the body: exteroceptors, visceroceptors (also called interoceptors), and proprioceptors.

Critical Thinking Questions

1. Based on Model 1, list the three classes of receptors in the space below.

2. Which of these does your group think is located in the external portions of your body?

3. Which of these does your group think is located in the internal portions of your body?

4. Close your eyes and touch your fingertip to your nose with your eyes closed. Reach behind you and touch your right and left hands together. Congratulations! You just used your proprioceptors. Discuss this phenomenon with your group, and then write, as best as you are able, a group description of what a proprioceptor does.

Application

5.   Classify each of the following as an exteroceptor, visceroceptor, or prioprioceptor.

a. Osmoceptors (measure osmotic pressure of blood) in hypothalamus:

b. Phasic receptors in your muscles that let you know if a body part is moving:

c. Stretch receptors in your bladder (let you know it's time to go to the bathroom):

d. Photoreceptors in the eye (let you perceive your external environment):

e. Chemoceptors in the liver that measure blood glucose levels:

f. Tonic receptors (tell you where your body is located in space):

g. Thermoceptors (detect temperature near the skin):

h. Nociceptors detect pain [trick question]:

Model 2: Classification by Stimulus

There are six classes of stimuli that receptors can detect.

Receptor Type	Stimulus Detected	Example
Mechanoreceptor	Any mechanical stimulus that will deform or change the position of the receptor.	
Chemoreceptor		CO_2 detectors in medulla and pons.
Thermoreceptor		
Nociceptor		Nociceptors (there aren't multiple examples of these)
Photoreceptor	Light (photons)	
Osmoreceptor		Hypothalamic thirst center

Critical Thinking Questions

6.   Using your knowledge of word roots, and the information in Question 5, finish filling in the table above.

7.   Classify each of these six classes as either exterocepors, visceroceptors, or both.

8.   Which of the six classifications in the table likely responds to anything that damages tissue?

9.   There are two types of nociceptors: **fast (A) pain fibers** and **chronic (B) pain fibers**.

    a. If you slam your finger in a car door, which of these two is likely being stimulated?

    b. Would that make this type of pain fiber an exteroceotor or interoceptor?

    c. Type B fibers detect visceral pain (e.g. gall stones). Would they be considered extero- or interoceptors?

Model 3: Classification by Structure

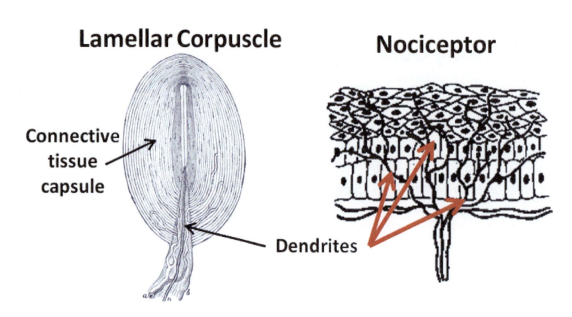

10.   Based on the model, what do both of these receptors have in common?

11.   What is different about their structures?

12.   Which one would you classify as an **encapsulated nerve ending**?

13.   Which one then is a **free nerve ending**?

Application

14. Golgi tendon receptors are proprioceptors that are buried in the tendons of skeletal muscles.

   a. What tissue type are tendons made from?

   b. If Golgi tendon receptors are <u>buried</u> in a tendon, based on your answer to 14a, are they encapsulated or free nerve endings? Be able to justify your answer to the class.

Endocrine Glands and Hormones

"What are endocrine glands and what do they make?"

Model 1: Development of Glands

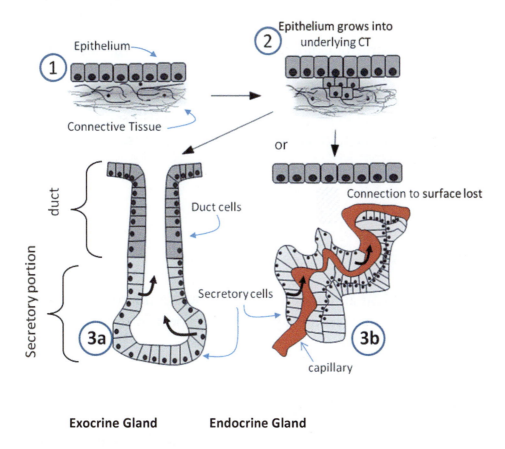

Exocrine Gland   Endocrine Gland

Critical Thinking Questions

1. Look at the shape of the epithelial cells vs the connective tissue in Part 1 of the model. Compare that to what remains in the mature exocrine gland in Part 3a. Which tissue from Part 1 is actually used to construct the exocrine gland in Part 3a?

2. Which tissue is used to construct the endocrine gland (Part 3b)?

3. Based on the model, what are the 2 major Parts or components of an exocrine gland?

4.   Where do the secretory cells of an exocrine gland secrete their products?

5.   Where do the secretory cells of endocrine glands secrete their products? Is this different from exocrine glands?

6.   What is the major structural difference between an endocrine and exocrine gland?

7.   As a group compose a grammatically correct English sentence that defines the difference between endocrine and exocrine glands.

Application

8.   If you were to look at an endocrine gland under a microscope, what would be the two main structural features you would expect to see?

Model 2: Major Endocrine Glands

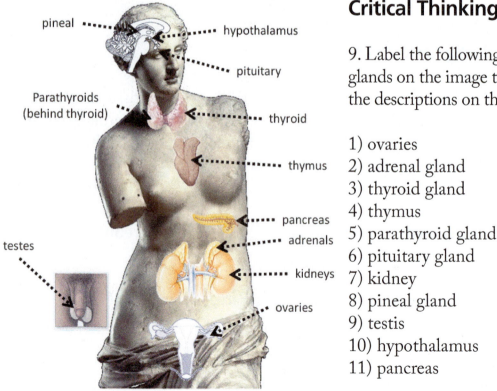

Critical Thinking Questions

9. Label the following endocrine glands on the image to the left using the descriptions on the next page.

1) ovaries
2) adrenal gland
3) thyroid gland
4) thymus
5) parathyroid gland
6) pituitary gland
7) kidney
8) pineal gland
9) testis
10) hypothalamus
11) pancreas

Definitions:

Ovaries – found in females, located in pelvic cavity

Adrenal glands – pyramidal glands on top of kidneys

Thyroid gland – bi-lobed gland surrounding trachea

Thymus – bi-lobed gland located inferior to thyroid and anterior to heart

Parathryoid glands – 4 small glands located within the thyroid

Pituitary – dangle d from the base of the mid-brain

Kidneys – bean-shaped organs on lateral abdominal walls

Pineal gland – posterior to the diencephalon of the brain, near corpus callosum

Testis – male sexual glands, located inside scrotum

Hypothalamus – Part of diencephalon of brain, superior to pituitary

Pancreas – located in the medial portion of the abdominal cavity.

Model 3: The Chemical Structures of Some Hormones

Aldosterone

Norepinephrine

Parathyroid hormone*

Oxytocin*

Estradiol

Cortisol

*Oxytocin contains 8 amino acids, Parathyroid hormone contains 84 amino acids.

Critical Thinking Questions

10. Just looking at the structures in Model 3, do any of these hormones look similar to one another?

   a. Which ones look similar?

   b. What is the same about all of them?

11.   Think back to the major classes of biological molecules you learned
       about at the beginning of A&P and look at the hormones your group
       said were similar. What is the term that describes the general structure of
       these hormones? [Librarians, you may look this up if you need to]

12.   Amino acids, the building blocks of proteins are usually small molecules
       with a terminal amine group. Peptides are short chains of amino acids
       (2 - 20) and proteins are long chains of amino acids (> 20)

       a. Which of the remaining hormones is a protein?

       b. Which one is a peptide?

       c. Which is a modified amino acid?

13.   As a group write a grammatically correct English sentence differentiating
       steroid hormones from non-steroid hormones.

Exercises

1.   Exocrine glands differ from endocrine glands primarily because exocrine
      glands have _____.

2.   Endocrine glands secrete substances called _____directly
      into the _____.

3.   Hormones can be broadly classified as either _____
      hormones which are hydrophobic or _____ - _____
      hormones which are hydrophilic.

4. List the hormones produced by the following glands, and indicate if they are steroid or non-steroid hormones. (Use the back of the page if you need to)

Anterior pituitary

Posterior pituitary

Thyroid

Parathyroid

Thymus

Pancreas

Adrenal gland

Pineal gland

Ovaries

Testes

5-8. Classify the following hormones as either steroid or non-steroid hormones:

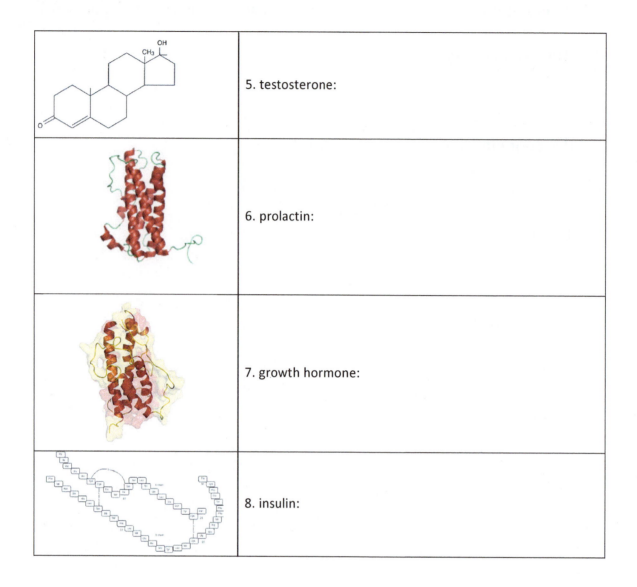

	5. testosterone:
	6. prolactin:
	7. growth hormone:
	8. insulin:

Hormone Mechanism of Action

"How do hormones exert their influence on cells?"

Model 1: Mechanism of Action of Steroid Hormones

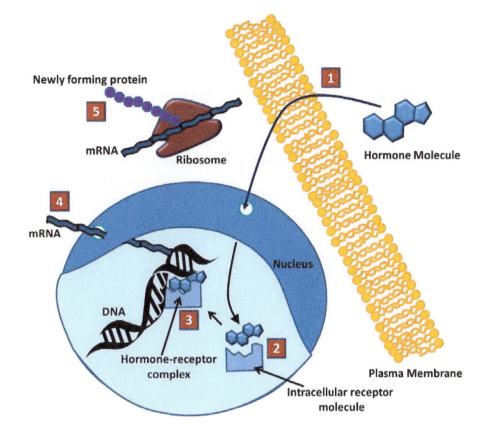

Critical Thinking Questions

1.   Which of the four main classes of biological molecules makes up the inner portion of cell membranes? (i.e. it's a _____ bilayer).

2.   Does this class of molecules mix well with water?

   a. What is our scientific word for things that dissolve in water? (It literally means "water loving" in Greek.)

   b. What is our scientific word for things that <u>don't</u> dissolve in water? (It literally means "water-fearing" in Greek.)

   c. Based on what we know about the chemistry of biological membranes, which of these two will be able to dissolve freely though the interior of a biological membrane?

3.   What class of hormone is represented in Model 1?
     Steroid

4.   Does it appear, based on the model, that this hormone has any problems getting into the cell?

     a. Based on this, would you say that steroid hormones are hydrophilic or hydrophobic?

5.   Where is the **receptor** for the steroid hormone located?

6.   What process does the **hormone-receptor complex** initiate in the nucleus of the cell? [If you can't remember the technical term, just describe the process]

7.   What type of molecule is eventually produced as a result of the steroid hormone entering the cell (cf #5 in the model).

     a. Was that molecule present before the steroid hormone entered the cell, or was it made new?

8.   In the space below, list some of the things that proteins are responsible for in human cells. Spokes-people: be prepared to share with the class.

Application

9.   Think about what kind of cells steroid hormones can enter, and explain why steroid hormones typically have systemic targets.

Model 2: Mechanism of Action of Non-Steroid Hormones

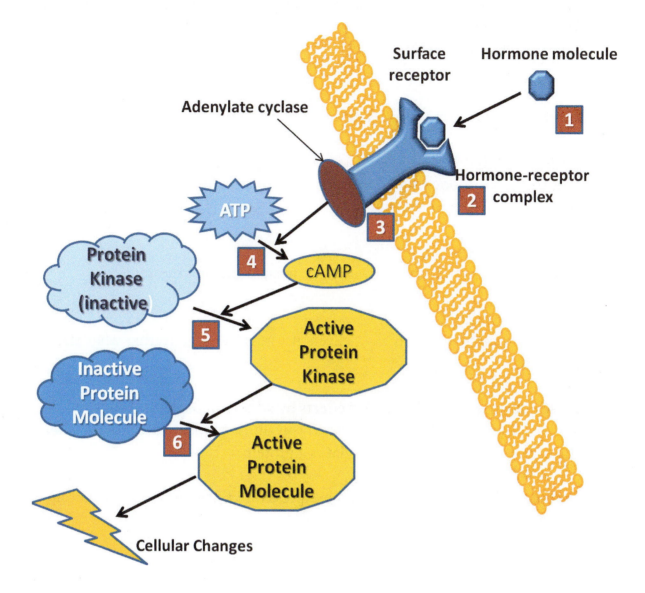

10. Look at the first step in the model above. Where is the receptor for the non-steroid hormone located?

a. Is this different from a steroid hormone? How?

b. What allows steroid hormones to pass through cell membranes?

c. How are non-steroid hormones different in t his respect?

11. The formation of the **hormone-receptor complex**, in this case, activates an enzyme anchored to the inside of the membrane. What enzyme is activated when the hormone binds its receptor? [Hint, remember that enzymes typically all have the same suffix]

12. This activated enzyme turns ATP into what?

13. Cyclic AMP (cAMP) is known as the <u>second messenger</u> because it actually delivers the hormone's signal to the cell cytoplasm. What delivered the message to the cell in the first place?

    a. So if cAMP is the second messenger, what is the first messenger?

14. The second messenger actually delivers its message by activating what class of enzymes?

15. These enzymes bring about their effects by adding phosphate groups to proteins. Based on the model, what do you think happens to a protein when one of these enzymes **phosphorylates** a protein?

16. Do non-steroid hormones cause the expression of <u>new</u> proteins like steroid hormones do?

    a. How then do they bring about their cellular effects?

Application

17. Precocious puberty is a disease caused by too much growth hormone being made too early. As a group come up with TWO drug targets that could be used to treat this disorder. [Hint: look at the model and think about what you could knock out to prevent GH from having any effect]

Exercise

18. Take a sheet of plain white paper. As a group write out a flowchart explaining the sequence of events involved in the mechanism of action of a non-steroid hormone on one half of the page, and the mechanism of action of a steroid hormone on the other half. Turn this page in.

Memorization fact: Although not true steroid hormones (they lack the cholesterol backbone) thyroid hormones, T3 and T4 are hydrophobic and exert their effects via the same mechanism as steroid hormones. These are the exception to the rule.

Regulation of the Endocrine System

"How are hormones controlled?"

Model 1: The Hypothalamic-Pituitary System

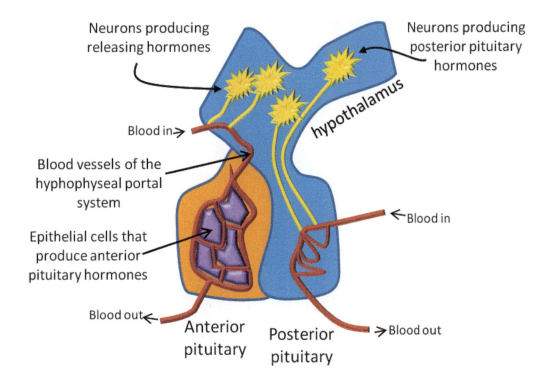

Critical Thinking Questions

1.   What type of cells secrete releasing hormones?

2.   Where are these cells located?

3.   What major organ is the hypothalamus part of?

4.   Based on the model, what circulatory structure do the hypothalamic releasing hormones use to get to the anterior pituitary?

POGIL
WWW.POGIL.ORG
Copyright © 2015

5.   What kind of cells secrete anterior pituitary hormones?

6.   Where are the cells that secrete anterior pituitary hormones located?

7.   What type of cells secrete posterior pituitary hormones?

8.   Are these cells actually located in the posterior pituitary gland?

   a. Where are they located?

   b. Does the posterior pituitary appear to produce any of its own hormones?

   c. How do the hormones get from the hypothalamus to the posterior pituitary to be released?

Application

9.   As a group, come up with a consensus answer to the following: Is the posterior pituitary actually an endocrine gland? Be able to support your answer based on evidence in the model.

Model 2: A Hypothalamic/Pituitary Negative Feedback Loop

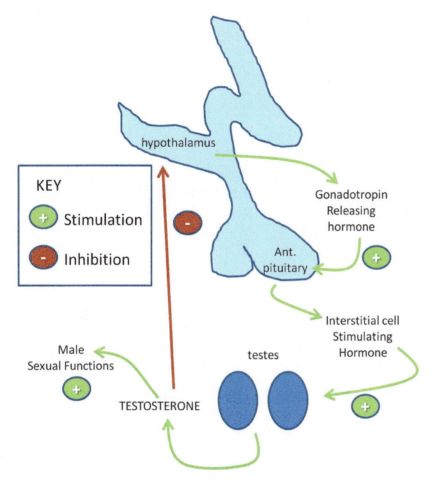

10. What hormone is being secreted by the hypothalamus here?

11. What is the target tissue for this hormone?

12. What does the hypothalamic hormone GnRH cause the ant. pituitary to do?

13. What is the target tissue for this hormone?

14.   What does ICSH cause the testes to do?

15.   What are the **TWO** functions of testosterone?

Application

16.   If increasing amounts of testosterone in the blood will *inhibit* the release of GnRH from the hypothalamus, what will promote the release of GnRH?

17.   The process outlined here is an example of a **negative feedback loop**. As a group come up with a definition, IN YOUR OWN WORDS, of what a negative feedback loop is.

18.   GnRH and ICSH are both examples of **tropic hormones**. Based on the model, what do these two hormones have in common?

a. Define 'tropic hormone'.

Model 3: Anterior Pituitary Hormones with their Targets

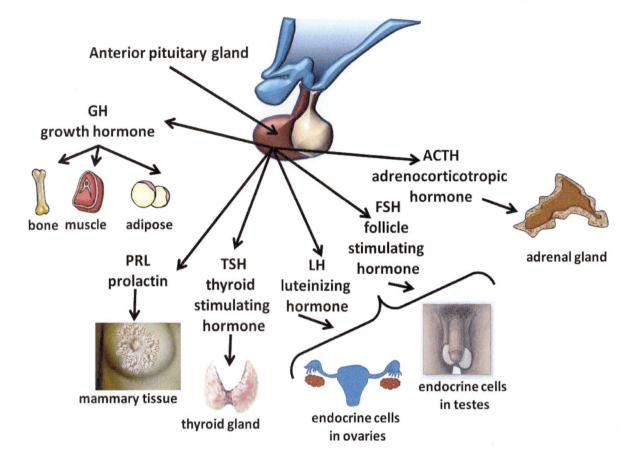

19. Using the model, list the six **anterior pituitary hormones** in the space below.

20. In women, FSH causes the ovaries to produce estrogen and LH causes the ovaries to produce progesterone. In men, FSH causes sperm cells to mature and LH stimulates the testes to produce testosterone. What is LH called in men?

21. Based on the names and target tissues of the six hormones you listed in CTQ 19, which ones are tropic hormones?

22. As a group, explain why the anterior pituitary is often referred to as the "master gland" of the endocrine system.

Exercises

Draw a negative feedback loop for the following sets of hormones:

1.   Hypothalamus: corticotropin releasing hormone
      Anterior Pituitary: adrenocorticotropic hormone
      Adrenal gland: cortisol

2.   Hypothalamus: Thyrotropin releasing hormone
      Anterior Pituitary: Thyroid stimulating hormone
      Thyroid gland: Thyroid hormones

3. Fill in the table below

Hormone (abbreviation)	Secreted by	enters the bloodstream @	target tissue
GH	Anterior Pituitary		
		Anterior Pituitary	Mammary glands
TSH			Thyroid Gland
	Anterior Pituitary		Adrenal cortex
LH			
			Germ cells in ovaries or testes
OT		Posterior pituitary	
	Hypothalamus		Kidneys

Red Blood Cells

"What are the cellular components of human blood and what do they do?"

Model 1: Hematocrit

Centrifuged blood sample

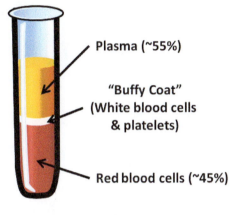

Plasma (~55%)

"Buffy Coat"
(White blood cells
& platelets)

Red blood cells (~45%)

Critical Thinking Questions

1. Blood can be separated in a centrifuge into plasma and formed elements. Based on the model (centrifuged blood sample), what are <u>formed elements</u>?

2. About what percentage of a typical blood sample is composed of white blood cells and platelets?

Model 2: Red blood cell morphology

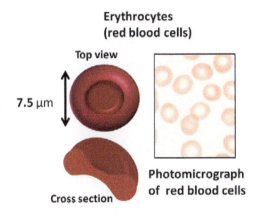

Erythrocytes
(red blood cells)

Top view

7.5 μm

Cross section

Photomicrograph
of red blood cells

3. As a group come up with a description of the shape of a red blood cell (<u>erythrocyte</u>).

4. What flying-disc toy do red blood cells resemble?

a. If you had one of those that
it into a 6" pipe, how would you

a. If you had one of those that was 7.5" in diameter and you wanted to get it into a 6" pipe, how would you do that?

b. Some capillaries are <6 μm in diameter. Given the answer to the previous questions, explain why erythrocytes might be shaped the way they are.

POGIL
WWW.POGIL.ORG
Copyright © 2015

Memorization fact: this shape also increases the surface area of the cell and makes them more efficient at gas exchange.

Model 3: Regulation of erythrocyte production

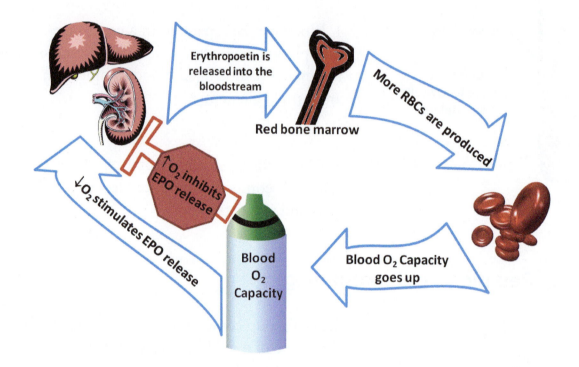

Critical Thinking Questions

5.   Erythropoietin is secreted by glandular epithelial cells **directly into the bloodstream**. What type of biological signaling substance is erythropoietin?

6.   Based on the model, where do erythrocytes come from?

7.   What **type** of control mechanism regulates the number of erythrocytes in the blood (Hint: we talked about these a lot with hormones)?

8. What is the relationship between erythrocytes and the oxygen-carrying capacity of the blood implied by the model?

9. What do erythrocytes do?

Application

10. Lance Armstrong was caught cheating in a Tour de France race because his blood showed abnormally high levels of erythropoietin. What advantage would an endurance athlete like a bicyclist have by taking extra erythropoietin?

Model 4: Recycling Eythrocytes

Critical Thinking Questions

11. According to the model, about how long do erythrocytes live?

12. a. Is there a nucleus inside the erythrocyte in Model 2?

b. Can a cell make mRNA without the DNA in the nucleus?

c. Can a cell make proteins without mRNA?

d. Why do red blood cells have a finite life-span?

13.   If you were to weigh an erythrocyte, about a third of its weight would be the protein hemoglobin.

a. What is the function of an erythrocyte?

b. Since erythrocytes have 100X more hemoglobin than any other protein, and knowing the function of an erythrocyte, <u>what is the function of hemoglobin</u>?

14.   As a group, write out a flowchart outlining the process by which hemoglobin is recycled. Note that bilirubin and biliverdin are the only components that are removed as waste (bile pigments are part of feces), so don't forget to account for the globin and the iron.

Application

15.   Vegans sometimes suffer from a form of <u>anemia</u> (lowered oxygen carrying capacity of the blood). What nutrients are they likely lacking from their diet?

Exercise

1.   Write out a flowchart or use a paragraph to describe the life cycle of an erythrocyte starting from the body's need to make it to its eventual destruction and recycling of its contents. You may use your textbook to check your answer, but not to formulate it.

      You should use the back of the paper.

ABO and Rh Blood Groups

"Why can we donate blood to some people but not to others?"

Model 1: Blood Surface Antigens and Plasma Antibodies

Type A	Type B	Type AB	Type O
Surface antigen A	Surface antigen B	Surface antigens A & B	Neither A or B antigens
Anti-B antibodies	Anti-A antibodies	Neither anti-A or anti-B antibodies	Both anti-A and anti-B antibodies

Critical Thinking Questions

1. If Dr. Brown has type A blood, what cell-surface marker proteins or **antigens** does he have on his red blood cells (RBC)?

2. Looking at the bottom half of the model, what antibodies are likely present in Dr. Brown's blood?

3. Dr. Mrs. Brown has type O blood, according to the model what surface antigens does she have on her red blood cells?

4. What antibodies are likely present in her blood?

POGIL
WWW.POGIL.ORG
Copyright © 2015

Model 2: ABO Mis-match Reaction

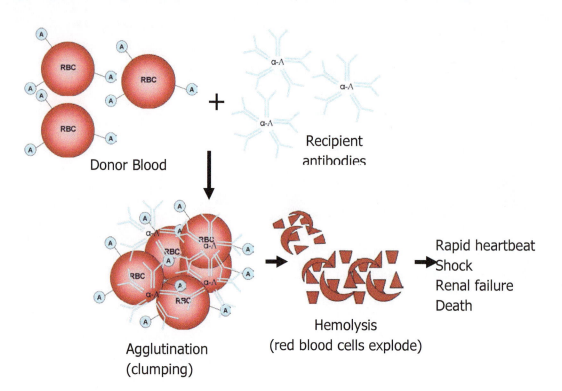

Donor Blood

Recipient antibodies

Agglutination (clumping)

Hemolysis (red blood cells explode)

Rapid heartbeat
Shock
Renal failure
Death

Critical Thinking Questions

5.   What is the blood type of the donor?

6.   What is the blood type of the recipient?

7.   What reaction occurs between the donor's red blood cells and the recipient's opposing antibodies?

   a. What process follows agglutination?

8.   Blood mis-matches can result in a condition called **Acute Hemolytic Reaction**. What are the symptoms of this reaction?

9.   Would agglutination have occurred if the recipient was given type O blood cells?

   a. Provide a consensus explanation for your answer.

Memorization fact: In addition to ABO, there is another component of blood type – the **Rh factor**. People who posses Rh antigens are referred to as Rh positive (e.g. O+ have nether A or B but they do have Rh), people without Rh antigens are Rh negative (A- would have the A antigen but not B or Rh).

Model 3: Erythroblastosis Faetalis
(a disease in which a pregnant woman's anti-Rh antobodies destroy fetal red blood cells and typically results in miscarriage).

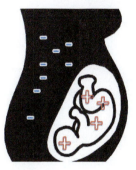 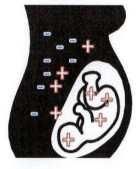  

Woman with RH negative blood type is pregnant with an Rh positive baby

Mother produces no anti-Rh antibodies at the beginning of pregnancy, but she is exposed to fetal Rh antigens

After being exposed to the Rh antigen, mother is now <u>sensitized</u> and produces anti-Rh antibodies

Any subsequent exposure results in the anti-Rh antibodies destroying all Rh-positive cells

Critical Thinking Questions

10. Dr. Brown has the blood type A negative. According to the model and the memorization fact, to what antigen does the word "negative" refer?

11. Compare the first pregnancy with the second, does it appear that the first baby was affected by the mother's immune system? What triggers the mother's sensitivity to the Rh antigen?

12. Based on the model, if Dr. Brown has never been exposed to the Rh antigen, what do you think will happen if he is transfused with A positive blood?

13. What will happen if he is transfused with A+ blood a second time?

Exercises

1.   Using the cartoons below, write out the complete blood type (ABO and Rh) for each picture indicated.

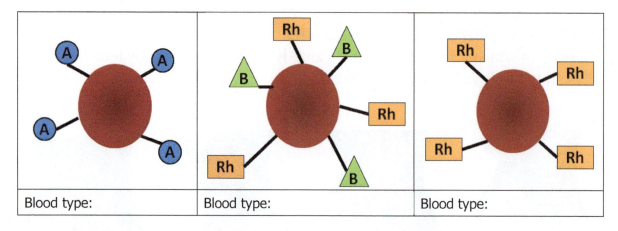

| Blood type: | Blood type: | Blood type: |

2.   Fill in the table for the indicated blood type, use an X to indicate if that particular presence is present in the indicated blood type (see example for Type A- blood):

Blood type	A antigens	anti-A antibodies	B antigens	anti-B antibodies	Rh antigens	Can receive a donation from
A-	X			X		A-, O-
A+						
B-						
B+						
AB-						
AB+						
O-						
O+						

3.   After filling in the table above, which blood type can be considered the universal donor?

4.   Likewise, which can be considered the universal recipient?

Hemostasis

"How does our body handle bleeding?"

Model 1: Blood Clotting

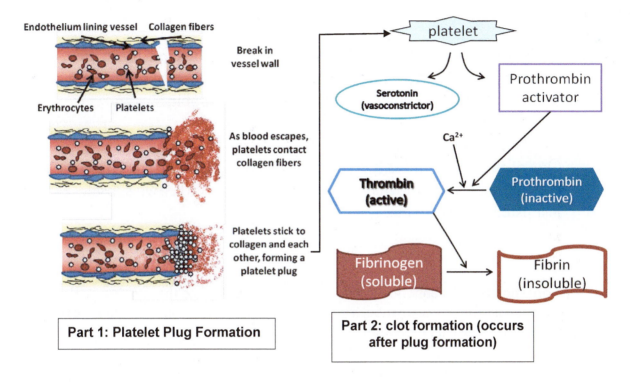

Part 1: Platelet Plug Formation

Part 2: clot formation (occurs after plug formation)

Critical Thinking Questions

1. Although not pictured in the model above, the first thing that happens when a blood vessel is torn is vasospasm, in which the walls of the blood vessel contract. As a group, come up with an explanation for this phenomenon.

2. Look at (Part 1 of) the model. Platelets are always present in the blood yet they don't (usually) form a plug. When they escape through a break in the blood vessel, what substance do they contact that induces them to become sticky and form a <u>platelet plug</u>?

3. According to Part 2 of the model, what two substances do platelets secrete after forming a platelet plug?

POGIL
WWW.POGIL.ORG
Copyright © 2015

4.   What is the inactive form of thrombin called?

5.   Why can't thrombin be in its active form in regular blood circulation? [What would happen if it was?]

6.   In the past you have learned about enzymes and cofactors. Both are necessary to convert prothrombin into thrombin. According to the model, what two substances convert prothrombin to thrombin?

7.   One of those substances is an enzyme, the other is a cofactor. Based solely on your knowledge of what an enzyme is, which of the two substances you just listed is the enzyme?

Memorization fact: An abnormal clot is called a **thrombus**, once that clot leaves the site of formation and enters circulation, it's called an **embolus**. If an embolus gets lodged somewhere it's called an **embolism**

Application

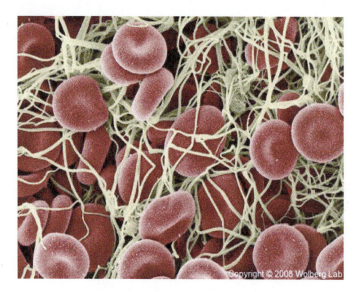

8. What is the final product in clot formation, according to Model 1?

The picture to the left is an electron micrograph of a blood clot. Based on your answer to the previous question, what is the stringy substance visible in this image?

9. Tissue Plasminogen Activator (TPA), Heparin, and Warfarin are all fibrinolytic substances. Think about what the suffix "-lytic" means and formulate an hypothesis as a group about what these compounds do.

Problems

1. Clinical biochemists are working on so-called "liquid bandages". These are liquids that can be applied to a wound to quickly stop blood loss. Circle the substances below that would be useful in one of these "liquid bandages"

PA	TPA	Prothrombin	Heparin	Thrombin
Warfarin	Fibrin	Fibrinogen	Calcium	Albumin

2. What are the common names for a cerebral embolism and a coronary embolism respectively?

3. Look at the list of compounds above. Which of those would you give a patient suffering from either of these conditions?

The Cardiac Cycle, Part 1

"What does it take to make a heartbeat?"

Model 1: The Cardiac Cycle

Semilunar valve

Anterior vena cava

Posterior vena cava

Aorta

Pulmonary artery

Right and left Atrium

Pulmonary veins

Antrioventicular valve

Right and left ventricles

Ventricles in diastole

Ventricles in systole

Critical Thinking Questions

1. Compare the two figures in the model. Are the ventricles of the heart larger during systole or diastole?

2. Everyone in the group make this shape with your hand. Now clench it into a fist. Imagine your hand is a chamber of the heart - Which is similar to diastole, the open shape or the clenched fist?

Which is more similar to systole?

POGIL
WWW.POGIL.ORG
Copyright © 2015

3.   In the space below list the five different tissues found in the human body.

4.   The heart is composed of three layers, the inner layer or *endocardium* (epithelium and connective tissue), the *myocardium*, and the outer *epicardium*.

     Look at the prefix "*myo–*" and decide as a group which of the tissues you listed above comprises the myocardium [Hint which one has myofibers, and myosin filaments, and myocytes?].

5.   What is different about this type of tissue in the heart? Does it have a special name? [Librarians - if your group is stuck you may consult the textbook]

6.   Cardiac muscle cells (**myocytes**) are connected to one another by gap junctions in the intercalated disks. These gap junctions allow ions to diffuse from one cell into all the adjacent cells.

     a. If a cardiac myocyte is brought to threshold (depolarizes), does it affect the cells around it? How so?

7.   Based on your answer to number 6, what does your group think will happen if a single myocyte in the atria of the heart depolarizes?

Model 2: The Cardiac Conduction System

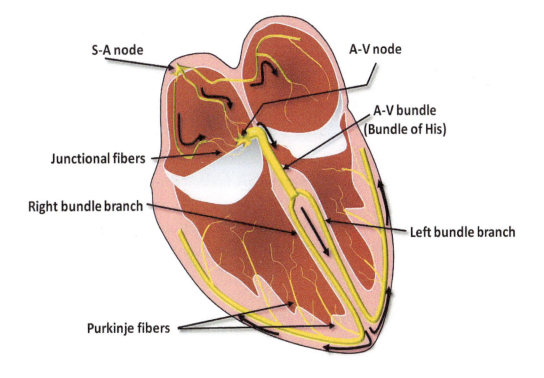

Fact Box: The heart contains special cardiac myocytes that don't have many contractile fibers in them. These cells are specialized to conduct cardiac muscle impulses throughout the heart muscle tissue - similar to nerves. These specialized cardiac myofibers are collectively called the **cardiac conduction system**. Like nerves, the thinner CCS fibers conduct an impulse more slowly than thicker fibers.

8.   Look at the model. The arrows indicate the pathway that a cardiac impulse will travel through the fibers of the CCS (in yellow). Follow the arrows backwards - where does the impulse begin?

9.   The cells of the SA node don't allow K+ ions to leave the cell as rapidly as most excitable cells. Use your knowledge of polarized cells to answer the following:

a) will these cells be able to repolarize as quickly as other excitable cells?

b) if a cell doesn't repolarize all the way, is it left closer to or further from the threshold voltage?

c) will this make them more or less likely to reach threshold spontaneously?

Librarians - look up the term *functional syncytium*. **Managers** - make sure everyone in your group can <u>explain</u> this term if called upon. **Everyone** - write out a definition in your own words below that you can share with the class.

---

10.   As an impulse travels along the fibers leaving the SA node, it contacts atrial myocytes.

a) What will happen to the myocytes that come into contact with these fibers?

b) What will happen to the myocytes that are touching the myocytes that are touching these fibers?

c) What will happen to the myocytes that are touching the myocytes that are touching the myocytes that are touching these fibers?

11.   As an impulse leaves the SA node, the myocytes of the atrial syncytium will depolarize. What happens to muscle cells when they depolarize?

a) Since all the myocytes in the atria are contracting (almost) at once, what will happen to the size of the atria?

12.   Will the connective tissue that separates the atrial syncytium from the ventricular syncytium conduct an impulse?

a) Then how will the impulse get to the ventricles?

13.   Look at the *junctional* fibers.

a) What effect will the diameter of the fibers have as the impulse travels toward the AV node?

b) What would happen if the impulse were to travel to the AV node before it had propagated throughout the atrial syncytium?

c) As a group, devise a consensus explanation for why the junctional fibers are the diameter they are.

14.   Look at the AV bundle and bundle branches.

a) Will an impulse travel relatively slowly or quickly along these fibers? Why?

b) Where do the purkinje fibers branch first in the ventricular syncytium?

c) Which end of the ventricles will begin to contract first?

Exercises

If you are still in class go ahead and start on these. If you need to work on these at home, take a transfer pipette with you.

1.   Fill your pipette bulb with water. You will have to invert the pipette and shake the water down into the bulb.

Now, holding the pipette with the bulb down -squeeze the bulb starting at the end closest to the neck. Make a note or use a marker to denote how far up the pipette the water travels.

Shake the water back down into the bulb.

Now squeeze the bulb starting at the end farthest from the neck and squeezing toward the neck. How far up the pipette does the water travel this time?

2.   Explain why the anatomy of the ventricular CCS is optimal for emptying the ventricles.

3.   Ventricular fibrillation is the uncoordinated contraction of cardiac myocytes. Explain why "shocking" the heart (think an ER doctor shouting "CLEAR!") can correct this problem.

4.   Sometimes an infection might damage a portion of the CCS requiring the implantation of an artificial pacemaker. What portion of the CCS is this artificial pacemaker replacing?

5. Place the following events in the cardiac cycle in order. Some events might occur simultaneously, put those side-by-side.

AV node fires	atrial systole	SA node fires
atria fill with blood	ventricular syncytium depolarizes	cardiac impulse travels down the bundle branches
cardiac impulse travels down the AV bundle	cardiac impulse travels along purkinje fibers	ventricular systole
atrial diastole (and repolarization)	atrial syncytium depolarizes	ventricles become filled with blood
ventricular diastole (and repolarization)	ventricles empty	

The Cardiac Cycle, Part 2

"What makes a heartbeat?"

Model 1: Monitoring the Electrical Activity of the Heart

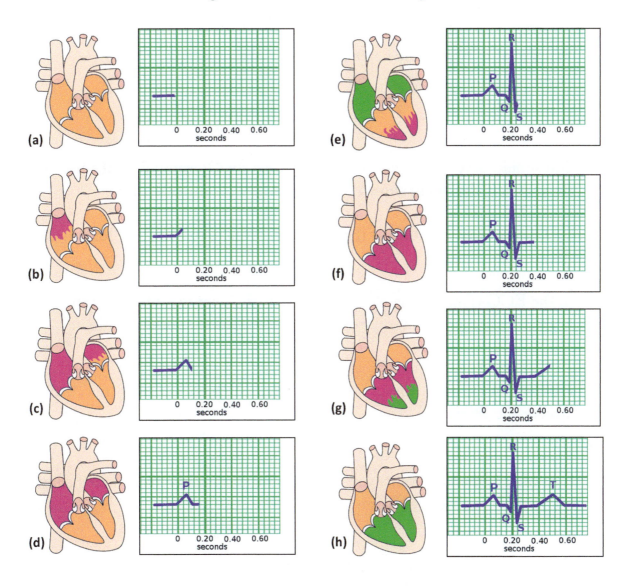

Purple represents **depolarization** of the cardiac myofibers. Green represents **repolarization** of the cardiac myofibers. Next to each heart image is a display of what you would expect to see on an ECG during these electrical events in the heart.

Critical Thinking Questions

1.   According to the model, what is electrical event is taking place in the heart when the P wave is being recorded on the ECG?

2.   What is occurring in the heart electrically when the QRS complex is being recorded?

3.   You should have listed two things as answers for Question 2. **Only one of those is actually being shown on the ECG.**

a. How does the amplitude of the P wave compare to that of the QRS complex?

b. The ECG is measuring electrical activity in one of the 3 layers of the heart, which one? [Hint: only one layer is composed of excitable tissue]

c. More muscle means more electrical activity. Which has more muscle the atria or the ventricles?

d. Which of the two events you listed in Question 2, is actually being shown on the ECG as the QRS complex?

4.   As a group, explain why the QRS complex is so much larger than the P wave.

5.   According to the model, what is occurring in the heart when the T wave is recorded.

Application

6.   Why isn't the repolarization of the atria visible on the ECG?

7.   Sometimes an inflammation of the heart (endocarditis) can damage the AV bundle or one of the bundle branches.

a. Will the ventricles still contract if one of the bundle branches is impaired (called a **bundle branch block**)?

b. Will the cardiac impulse travel through the ventricular syncitium as fast as normal?

c. Which part of the ECG will be affected by this? What might you expect to happen on the ECG?

Model 2: The Complete Cardiac Cycle

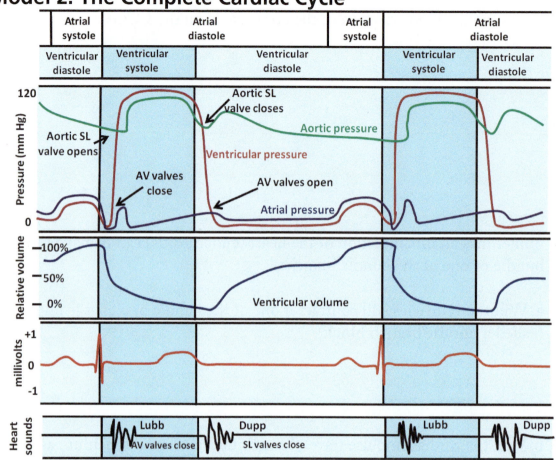

Critical Thinking Questions

8.   Look at the pressure graph.

   a. When the AV valves open, is pressure higher in the atria or ventricles?

   b. What about when the SL valves open?

   c. Do the AV valves open only during atrial systole?

9.   Look at the ventricular volume graph. What percentage (roughly) of ventricular volume is filled passively (i.e. without atrial systole)?

10. The <u>quiescent period</u> of the cardiac cycle refers to the period when both syncitia are in diastole. What is occurring in the heart during the quiescent period? [Hint: look at ventricular volume]

11. According to the model, what are the two main heart sounds?

    a. What causes each one?

Application

12. There are many ways to describe a single heartbeat or <u>cardiac cycle</u>. Examine the model and as a group define a cardiac cycle. When you are prompted, the instructor will have the reporter for each group write your response on the board. The reporter should be able to verbally explain your definition. There can be multiple correct answers based on the criteria your group chooses to use when defining a cardiac cycle.

Problems

1.   Atrial fibrillation is the rapid uncoordinated contraction of the atrial myocardium. In the space below, draw an ECG trace that would be characteristic of a atrial fibrillation.

2.   Based on what you know about the actions of the heart, explain why atrial fibrillation is not considered life-threatening.

3.   Likewise, explain why ventricular fibrillation is life-threatening and requires the use of a cardiac defibrillator.

Capillary Exchange

"How does stuff get from your blood to the tissues?"

Model 1: Systolic and Diastolic Pressure in Major Blood Vessels

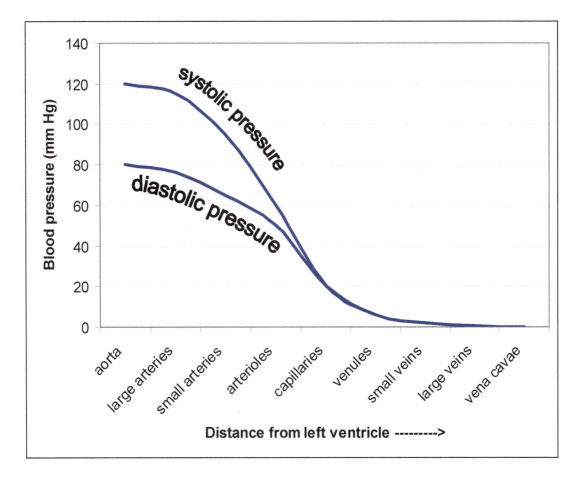

Critical Thinking Questions

1.   Based on the model, where is blood pressure the highest?

2.   The systolic and diastolic labels refer to the pressure in these blood vessels when one half of the heart in particular is in either systole or diastole. Which heart chambers does your group think the systolic and diastolic labels refer to?

Application

3.   Why isn't someone's blood pressure usually measured from a vein?

Model 2: Blood Components and their sizes*

formed elements (7.5 - 50 µm)	blood gases (<1 nm)	sugars (1 nm)
small proteins/antibodies (6 nm)	large proteins (albumins/fibrinogen) (8-10 nm)	water (< 1 nm)
peptide hormones (1-5 nm)	protein hormones (5-9 nm)	steroid hormones (3 nm)
electrolytes (< 1nm)	plasma wastes (<1 - 3 nm)	

Critical Thinking Questions

4.   One or more of the components listed above will be able to freely cross the capillary wall via diffusion *through the endothelial cell membrane*. Which one(s) is/are those? [Hint: think about hydrophobicity]

5.   Many capillaries are <u>continuous</u>, that is the cells that form their walls are pressed against each other, leaving only a tiny ($\approx$ 1nm) slit. The individual cells still contain channels for water and ions. Which of the components above will be able to diffuse through these channels and slits?

6. Some of the components above are too large to fit through ion channels in the cells and the slits between the cells. Which ones are these?

7. If necessary, technicians may consult their textbooks to answer this question: By what process do cells transport large molecules or macromolecular Particles across their membranes?

Model 3: Exchange in the Capillaries

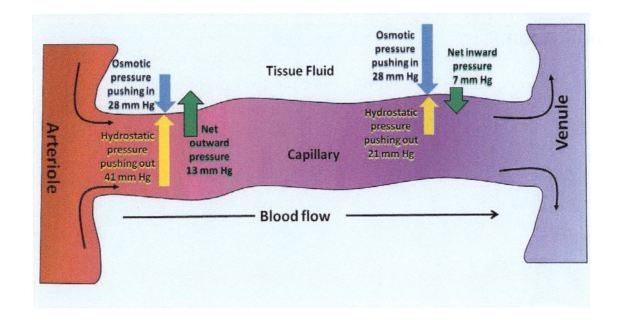

Critical Thinking Questions

8. According to the model, which is greater at the arterial end of the capillary - hydrostatic pressure or osmotic pressure?

a. In which direction (out of the capillary or into the capillary) will fluid travel at the arterial end of the capillary?

9.   Which is greater at the venular end of the capillary - hydrostatic pressure or osmotic pressure?

a. In which direction will fluid travel at the venular end of the capillary?

10.   Provide a consensus answer and explanation to this question: Does all the fluid that leaves the blood at the arterial end return at the venous end?

Fact Box: Due to the high hydrostatic pressure at the arterial end of a capillary, not only are liquids and small dissolved molecules squeezed out, but many small proteins too. The high hydrostatic pressure actually pushes on the walls of the capillary so hard that larger gaps open up at the seams between cells!

Application

11.   As a group define the following three mechanisms of capillary exchange and provide a brief (one sentence) explanation of the mechanism of each method.

a. Diffusion

b. Endo/exocytosis

c. Filtration (bulk flow)

12.   Which plasma component is responsible for maintaining the osmotic pressure in the plasma? [Hint: it needs to be too large to leave the capillary via bulk flow, but small enough to still be dissolved]

Hemodynamics

"What factors affect blood pressure?"

Model 1: Cardiology Patient Records for Al Fields

Cardiac Health Assessment: Conducted 09/18/1979		
Observations: Patient in good health, age 22, height 178 cm, weight 82 kg		
Resting Values	**After walking for 1 min**	**After walking for 10 min**
HR: 70 bpm	HR: 75 bpm	HR: 85 bpm
SV: 0.071 Lpb	SV: 0.071 Lpb	SV: 0.079 Lpb
CO: 5.0 Lpm	CO: 5.5 Lpm	CO: 6.1 Lpm
BV: 5.0 L	BV: 5.0 L	BV: 5.0 L
PR: 18.66 mmHg/L/min	PR: 18.66 mmHg/L/min	PR: 18.66 mmHg/L/min
BP: 120/80 mmHg	BP: 138/85 mmHg	BP: 162/90 mmHg

Patient Hospitalization Records		
Date: 01/01/1981	**Date:** 11/27/2007	HR (heart rate: beats per minute)
Diag: MVA w/ hemorrhage	**Diag:** Atherosclerosis	SV (stroke volume: liters per beat)
HR: 70 bpm	HR: 70 bpm	CO (cardiac output: liters per minute)
SV: 0.071 Lpb	SV: 0.071 Lpm	BV (total blood volume: liters)
CO: 5.5 Lpm	CO: 5.0 Lpm	PR (peripheral resistance)
BV: 4.2 L	BV: 5.0 L	BP (blood pressure)
PR: 13.22 mmHg/L/min	PR: 21.33 mmHg/L/min	
BP: 90/64 mmHg	BP: 140/90	

Critical Thinking Questions

1. Examine the health assessment data for Mr. Fields:

    a. What is the mathematical relationship between heart rate and blood pressure? [i.e. when one goes up, what does the other do?]

    b. What is the mathematical relationship between stroke volume and blood pressure?

POGIL
WWW.POGIL.ORG
Copyright © 2015

2.   In the space below write the **units** for heart rate, stroke volume, and cardiac output **in the designated space**. Write them as fractions, e.g. heart rate would be "beat/minute."

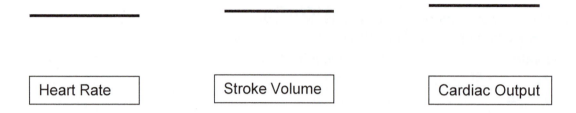

Heart Rate   Stroke Volume   Cardiac Output

Just looking at the fractions you've written on the previous page, insert mathematical operators (plus signs, multiplication signs, equals sign) into your units above so that you devise an <u>equation</u> that explains the relationship between heart rate, stroke volume, and cardiac output. i.e. <u>make the units cancel out</u>!

3.   Write a consensus definition of cardiac output. Be able to explain your definition to the class.

4.   What is the relationship between cardiac output and blood pressure?

5.   There are three factors that influence blood pressure. We have established that cardiac output (which includes stroke volume and heart rate) is one. Based on the data, what are the other two?

Application

6.   When Al Fields was admitted in 1981, the first thing the ER team did was push 1 liter of IV buffered Ringer's solution into his circulatory system. Why?

7.   Atherosclerosis is the narrowing of the arteries due to build up of plaques on the artery walls. Explain why people with atherosclerosis have elevated blood pressure.

Memorization fact: The best means of understanding total hemodynamics is the Mean Arterial Pressure (MAP). MAP is defined as diastolic pressure plus pulse pressure divide by three. Pulse pressure is the difference between the systolic and diastolic pressures. This can be expressed mathematically as:

$$MAP = P_D + \frac{(P_S - P_D)}{3}$$

Where P_D is diastolic pressure and P_S is systolic pressure. This particular measure of hemodynamics is holistic because MAP is equal to the product of cardiac output and peripheral resistance (MAP = CO X PR), so it takes into account all of the factors that can influence blood pressure.

Model 2: Arteries of the Lower Limb

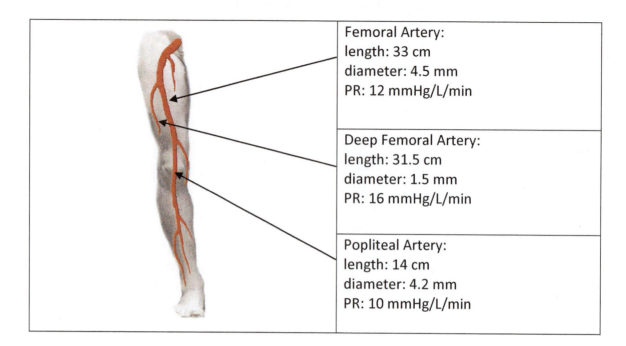

Femoral Artery:
length: 33 cm
diameter: 4.5 mm
PR: 12 mmHg/L/min

Deep Femoral Artery:
length: 31.5 cm
diameter: 1.5 mm
PR: 16 mmHg/L/min

Popliteal Artery:
length: 14 cm
diameter: 4.2 mm
PR: 10 mmHg/L/min

Critical Thinking Questions

8.   What is the main morphological difference between the femoral artery and the deep femoral artery, length or diameter?

a. Which has the higher peripheral resistance?

9.   What is the main morphological difference between the Femoral artery and the Popliteal artery?

a. Which has the higher peripheral resistance?

10.   What does your group think is the relationship between artery length, diameter, and peripheral resistance?

11. Is there a physiological mechanism that can adjust artery length significantly in order to maintain homeostatic blood pressure?

    a. Explain why vasodilation and vasoconstriction of the arteries (specifically arterioles) is the most effective means of maintaining blood pressure **from minute to minute**.[Make sure you discuss other mechanisms and why they might not always be effective]

Application

12. When a person donates blood plasma, the formed elements are returned to the body while the plasma is retained. If no extra fluids are added this will obviously cause a drop in blood volume. What effect if any will this have on blood viscosity?

13. How would this change in viscosity affect peripheral resistance? [Hint: think about trying to pour ketchup versus pizza sauce out of the same size bottle] What effect would this have on blood pressure?

Vocabulary (homework)

Use your textbook and what you have learned today to define the following terms in <u>your own words</u>:

Venous return

Preload

Afterload

End systolic volume

End diastolic volume

Frank-Starling Law of the Heart

Cardioinhibitor reflex

Cardioaccelerator reflex

Exercises (homework)

1.   At rest, Juliet's heart rate is 72 BPM and her CO is 5.0 LPM. When she sees Romeo her heart rate reaches 120 BPM and her CO is 15.0 LPM. What is her SV both before she sees Romeo and after?

2.   Mr. T has a BP of 120/80 what is his mean arterial pressure (MAP)? Assuming his CO = 5 LPM, what is the total peripheral resistance of Mr. T's arterioles?

   If Mr. T starts pitying fools and his peripheral resistance rises to 25 mmHg/L/min, what will his MAP be?

3.   Using the equations for MAP calculate MAP two ways for each of the boxes in Model 9A.

Cardiac Health Assessment: Conducted 09/18/1979		
Observations: Patient in good health, age 22, height 178 cm, weight 82 kg		
Resting Values	**After walking for 1 min**	**After walking for 10 min**
HR: 70 bpm	HR: 75 bpm	HR: 85 bpm
SV: 0.071 Lpb	SV: 0.071 Lpb	SV: 0.079 Lpb
CO: 5.0 Lpm	CO: 5.5 Lpm	CO: 6.1 Lpm
BV: 5.0 L	BV: 5.0 L	BV: 5.0 L
PR: 18.66 mmHg/L/min	PR: 18.66 mmHg/L/min	PR: 18.66 mmHg/L/min
BP: 120/80 mmHg	BP: 138/85 mmHg	BP: 162/90 mmHg
Patient Hospitalization Records		
Date: 01/01/1981	**Date:** 11/27/2007	HR (heart rate: beats per minute)
Diag: MVA w/ hemorrhage	**Diag:** Atherosclerosis	SV (stroke volume: liters per beat)
HR: 70 bpm	HR: 70 bpm	CO (cardiac output: liters per minute)
SV: 0.071 Lpb	SV: 0.071 Lpm	BV (total blood volume: liters)
CO: 5.5 Lpm	CO: 5.0 Lpm	PR (peripheral resistance)
BV: 4.2 L	BV: 5.0 L	BP (blood pressure)
PR: 13.22 mmHg/L/min	PR: 21.33 mmHg/L/min	
BP: 90/64 mmHg	BP: 140/90	

Problem

4.   Dr. Brown gets totally ripped over the summer and decides to try out for a triathlon. After taking some endurance and cardiology tests he gets the following results: Ventricular end systolic volume = 50 mL; ventricular end diastolic volume = 160 mL; HR = 140 BPM; and BP = 135/78.

Calculate the following for Dr. Brown: SV, MAP, CO, PR.

Innate Immunity

"How does the body prevent an infection from taking place?"

Model 1: A Schematic of the Body

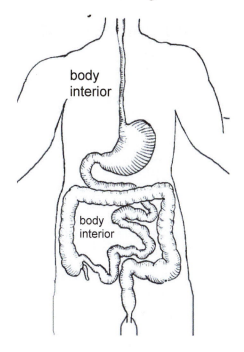

Critical Thinking Questions

1. Place your pencil tip somewhere in the esophagus. Lightly trace along the GI tract through the stomach, the intestines, and out the anus. Based on the drawing, did your pencil ever enter the actual interior of the body?

2. Based on the model, what would your pencil have to do in order to actually be in the body's interior?

3. Explain why objects inside the alimentary canal (gut cavity, gastrointestinal tract, coelom) are **not** technically in the body.

4.   Does the same apply for items in the lower urogenital or upper respiratory tract? Have they crossed the skin or a mucous membrane?

Model 2: Components of Innate Immunity

Fill in the first column below. You may all use your textbook or online resources - managers might wish to assign team members to a particular portion of the table.

Innate Immune Component	Definition
	Covers the entire body, largest organ
	Lines digestive, urinary, and reproductive tracts
	Lines respiratory tract, moves mucus + invaders out
	Protein catalysts that can harm or destroy pathogens nonspecifically. Examples include pepsin in stomach and lysozyme in tears
	Causes a very low pH environment, prevents growth of bacteria in upper portions of the digestive tract.
	Causes the surface of the skin to be very salty, preventing the growth of bacteria
	Proteins made and released by the body's cells in response to the presence of pathogens or tumors
	Cysteine-rich host defense peptides that are active against bacteria, viruses, and fungi
	Non specific lymphocytes that provide rapid responses to virally infected cells and that respond to tumor formation by inducing apoptosis in stressed cells
	A system comprised of a number of small proteins in the blood that form an activation cascade that activates the membrane attack complex
	The process by which cells engulf a solid particle, in the immune system it is a major mechanism used to remove pathogens and debris
	A tissue response to damage and/or pathogen invasion characterized by localized heat, pain, swelling, and redness
	A rise above homeostatic body temperature that inhibits bacterial growth and limits the availability of iron in the blood.

Critical Thinking Questions

5.   Some of the components that you filled in above are considered **barrier defenses** because they <u>prevent entry and attachment</u> of microbes. These are also called the *first line of defense* because they are the first obstructions a pathogen faces in trying to establish an infection. In the space below list all of the components from the first column above that would qualify as first-line or barrier defenses.

6.   Some barrier defenses are physical while some are chemical. Circle the physical barriers above.

7.   In the space below write those components from the first column that fight pathogens once they've penetrated the barriers. These are the second-line defenses.

8.   Do any of the components of innate immunity seem to be specific for any particular pathogen species? Based on that answer, write a definition (as a group) of innate immunity.

9.   Based on your understanding of barrier defenses, explain why large coverage burns are so often fatal.

Memorization Fact: There are two types of phagocytes in the blood: **neutrophils** and **monocytes**. When monocytes leave the bloodstream through the process of **diapedesis** and enter the surrounding tissues, they become **macrophages**.

Model 3: Inflammation

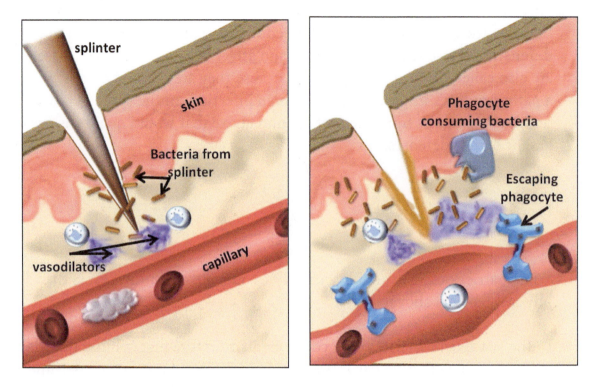

Critical Thinking Questions

10.   What is the difference between the blood vessel in the recently damaged
        tissue on the left and the inflamed tissue on the right?

        a. Why might this be beneficial?

11.   The four cardinal symptoms of inflammation are **calor, dolor, rubor,** and
        tumor. What do these mean in English? [Hint - look at the table in Model 2]

12.   Explain why three of these four symptoms can be attributed to increased
        blood flow.

Exercises

1.   Fill in the bottom of the chart below:

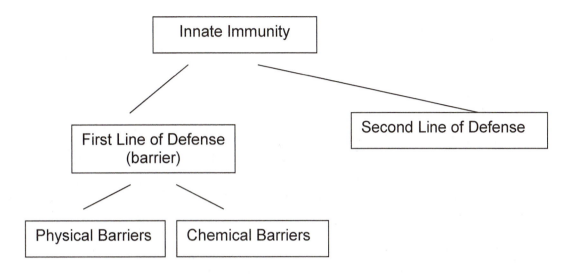

2.   What are the two ways that fever helps fight infection?

3.   Salivary amylase is a digestive enzyme found in saliva. How would you classify this immune component?

4.   How do vector-born diseases like malaria and west nile virus penetrate the barrier defenses?

5.   Vaginal pH is maintained by bacteria that live in the vaginal canal. Explain why removing these bacteria can result in greater risk for yeast infection.

Adaptive Immunity:
T-cells and the Cellular Immune Response

How does the body recognize and fight foreign invaders?

Model 1: T-cell Maturation and Self-inventory

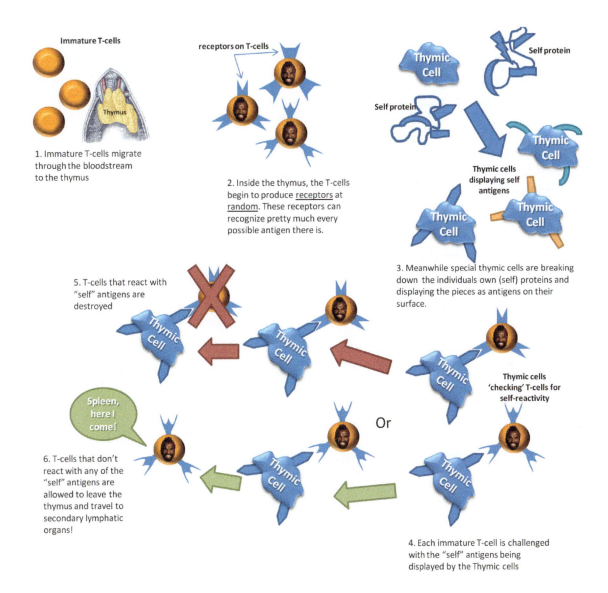

Immature T-cells

1. Immature T-cells migrate through the bloodstream to the thymus

receptors on T-cells

2. Inside the thymus, the T-cells begin to produce <u>receptors</u> at <u>random</u>. These receptors can recognize pretty much every possible antigen there is.

Self protein

Thymic Cell

Self protein

Thymic Cell

Thymic cells displaying self antigens

Thymic Cell

Thymic Cell

3. Meanwhile special thymic cells are breaking down the individuals own (self) proteins and displaying the pieces as antigens on their surface.

5. T-cells that react with "self" antigens are destroyed

Thymic Cell

Thymic Cell

Thymic Cell

Thymic cells 'checking' T-cells for self-reactivity

Spleen, here I come!

6. T-cells that don't react with any of the "self" antigens are allowed to leave the thymus and travel to secondary lymphatic organs!

Thymic Cell

Or

Thymic Cell

4. Each immature T-cell is challenged with the "self" antigens being displayed by the Thymic cells

Critical Thinking Questions

1.   T cells are lymphocytes, a type of leukocyte. Given this, where do these cells originate?

POGIL
WWW.POGIL.ORG
Copyright © 2015

2.   Based on where these cells mature, why do you think they are called T cells?

3.   Look at Part 2 of the model: with what antigens can the new T-cell receptors react with (bind)?

      a. Is it possible that some of these could react with your own proteins?

4.   Look at Step 3 of the model: where do the antigens displayed on the surface of the thymic cells come from?

      a. How many different antigens can a single T-cell recognize? [how many different **types** of receptor does each cell have?]

5.   Look at Step 5 of the model: what happens to T-cells that have receptors that bind to the "self" antigens from the thymic cells?

6.   Look at step 6 of the model: what happens to the T-cells that have receptors that don't bind to the "self" antigens on the surface of thymic cells?

7.   Knowing that T-cells are tasked with the job of destroying cells that they interact with, why is the **immune self-inventory** outlined in Model 1 so important?

Application

8.   Explain the possible outcomes if T-cells were able to react with "self" antigens [Hint: what if a T-cell that could recognize pancreatic beta cell antigens was allowed to survive?].

Model 2: T-cell Activation

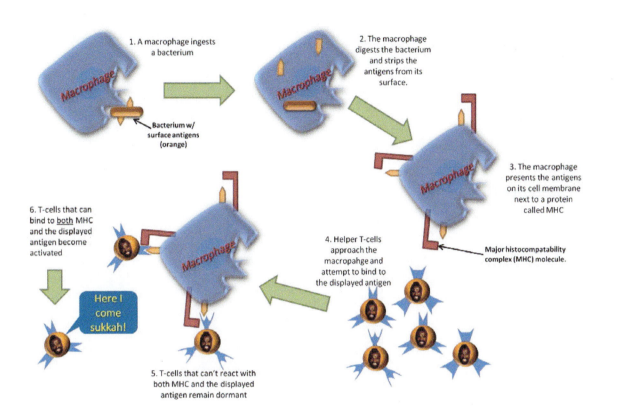

1. A macrophage ingests a bacterium

Bacterium w/ surface antigens (orange)

2. The macrophage digests the bacterium and strips the antigens from its surface.

3. The macrophage presents the antigens on its cell membrane next to a protein called MHC

Major histocompatability complex (MHC) molecule.

4. Helper T-cells approach the macropahge and attempt to bind to the displayed antigen

6. T-cells that can bind to both MHC and the displayed antigen become activated

Here I come sukkah!

5. T-cells that can't react with both MHC and the displayed antigen remain dormant

Critical Thinking Questions

9.   According to the model, what type of cell is the first to encounter a bacterial invader?

      a. What is the process of eating another cell called?

10.   Based on the model, why are certain macrophages called antigen-presenting cells (APCs)?

11.   How many <u>different</u> types of antigen-recognizing receptors does a single T-cell have? [you already answered this earlier]

12.   T-cells interact with their specific antigen AND a particular <u>macrophage surface protein</u>. What is the surface protein?

13. What two interactions are necessary for T-cell activation?

Memorization fact: MHC stands for Major Histocompatibility Complex.

Application

14. Explain the following sentence: MHC allows a T-cell to recognize only **presented** antigens.

15. Would it be a good thing to activate immune cells when they are not needed? Explain.

Model 3: Different T-cells Perform Different Functions

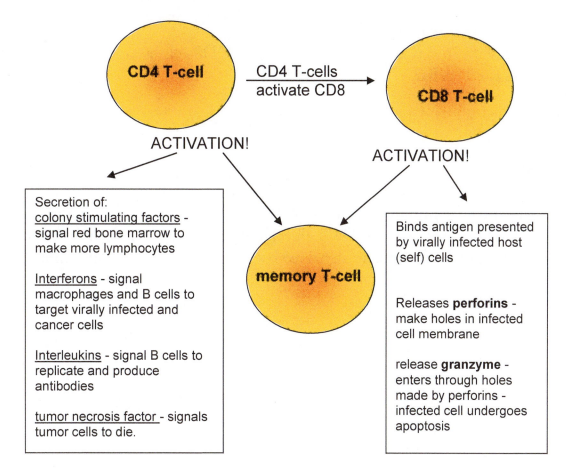

Critical Thinking Questions

16.   Based on the model, what are the three types of T-cells?

17.   CD4 T-cells secrete four different types of chemical factors. As a whole, what do these factors do? [Hint: they all do essentially the same thing, what is it?]

18.   Does it appear from the model, that CD4 T-cells actually destroy or damage foreign cells?

        a. Based on your answer above - why are CD4 T-cells referred to as **helper T-cells**?

19.   What is the fate of a cell that interacts with a CD8 T-cell?

       a. Based on your previous answer - why are CD8 cells also referred to as **cytotoxic T-cells**?

Exercises

1.   **Cytokines** are immune signaling molecules. What are the four major classes of cytokines secreted by CD4 (helper) T-cells, and what is their definition? Colony stimulating factors - signal red bone marrow to make more lymphocytes

2.   In the space below, explain, in your own words, the process involved in T-cell maturation.

3.   On a clean piece of paper, write out the process involved in the **cellular** (T-cell mediated) **immune response** starting with when a foreign invader gets into the body.

Adaptive Immunity:
The Humoral Immune Response

"Where do antibodies come from and what do they do?"

Model 1: B-cell Activation

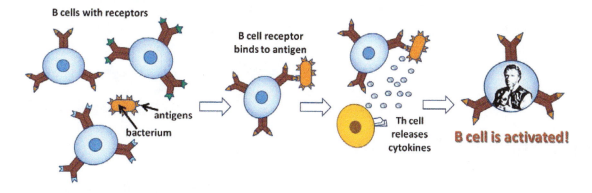

Critical Thinking Questions

1.   According to the model, only one of the three B-cells is activated. What was different about the one that got activated compared to the other two? [Hint: look at the receptors]

2.   Look carefully at the model. Did binding to the antigen automatically activate the B-cell?

   a. The B-cell had to interact with another type of immune cell, which one?

   b. Did that other immune cell just up and start activating the B-cell or was there a necessary preliminary step before it could start interacting with the B-cell?

   c. According to the model, what two steps are required for B-cell activation?

POGIL
WWW.POGIL.ORG
Copyright © 2015

Model 2: Post-activation Activities of B-cells

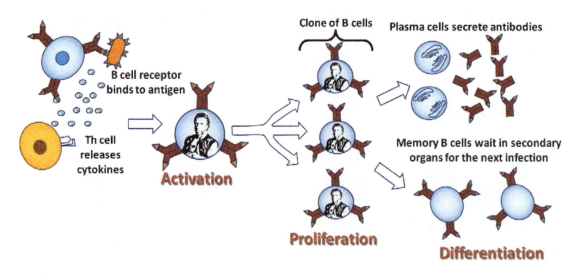

Critical Thinking Questions

3.   According to the model, what two things contribute to B-cell activation?

4.   What is the first step following activation?

5.   Look at the cells labeled "clone of B cells". Are the receptors on their surface the same or different from the parent cell?

   a. Are the receptors on the clone cells the same (as each other)?

   b. When you hear the word clone, does that indicate that the things are genetically identical, similar, or different?

   c. As a group, explain what is meant by the expression "clone of B-cells"

6.   What step in B-cell activation follows proliferation?

7.   What two types of cells does the B-cell differentiate into?

8.   What is the function of each cell type?

Model 3: Antibody Classes and their Functions

Antibody class	Abundance (in bloodstream)	Location	Functions
IgA	13%	blood and exocrine secretions (tears, saliva, breast milk)	▪ binds to bacteria and viruses, prevents them from attaching to host cells ▪ binds to toxins- inactivates them
IgD	<1%	surface of B-cells	▪ Is the same thing as the B-cell receptor
IgE	<1%	mostly in intestinal tract, some in circulation (that's why is <1%)	▪ Causes degranulation of mast cells and basophils ▪ mediates inflammatory response ▪ responsible for allergic hypersensitivity ▪ tags worms for eosinophil recognition
IgG	80%	primarily in blood, also in tissues and cavities	▪ bind viruses and prevent them from attaching to host cells ▪ bind to toxins and neutralize them ▪ agglutination of bacteria ▪ opsonization ▪ tag for NK cells to recognize ▪ activate complement
IgM	6%	blood and tissue fluid	▪ 1st antibody secreted ▪ agglutination of bacteria ▪ responsible for ABO mismatch reaction ▪ complement activation

Note: Memorize this table!!!!

Critical Thinking Questions

9.   After B-cell differentiation into plasma cells, what is the first antibody class to be secreted?

a. What are the three major functions of this antibody class?

b. Think back to what you know about blood plasma, what are the three types of plasma proteins?

c. Which ones are antibodies?

d. Since antibodies belong to that class of plasma proteins, and are responsible for immune functions, what do you think the Ig in IgM stands for?

10. Someone comes into your hospital emergency room with a severe allergic reaction to peanuts. When you get their blood work back, what antibody class would you expect to be elevated?

11.   Someone comes into your emergency room with a parasitic worm infection. Which antibody class would you expect to be elevated?

12.   Which Ig class is responsible for the immunity that is transferred from a mother to her nursing infant?

a. What are the functions of this Ig class?

13.   If someone has severe systemic bacteremia (bacteria in the blood, body cavities, and organs) which Ig classes should be elevated on their bloodwork?

14. Someone comes into the ER with a rattlesnake bite. You happen to have some rattlesnake anti-venom made from antibodies. Which Ig classes could this be made from?

Memorization Box: Passive immunity refers to the presence of antibodies in a person that were not produced by the person. **Active immunity** refers to the ability to produce antibodies against a particular antigen.

---

15. Based on the above definitions, what type of immunity was granted to the snake bite victim in CTQ 14?

    a. The rattlesnake anti-venom the patient received was anti-rattlesnake IgG that was produced in a horse. Would you consider this type of immunity artificial or natural?

16. Based on the above definitions, would a baby receiving IgA antibodies from its mother be passive or active? Natural or artifical?

17. A person receives an injection of killed polio viruses (a vaccine) which causes them to generate memory T-cells and B-cells, which confer lasting immunity to the virus. Is this passive or active immunity? Artificial or natural?

18. A three year-old gets the chicken pox, activating his immune system and causes the differentiation of memory T-cells and B-cells specific to the chicken pox virus. This child is no longer susceptible to the chicken pox virus. Does he have active or passive immunity? Artificial or natural?

Exercises

Define the following terms and give an example of each:

1.   Artificially acquired active immunity.

2.   Naturally acquired active immunity.

3.   Artificially acquired passive immunity.

4.   Naturally acquired passive immunity.

5.   Write out a flow chart that outlined the steps in the adaptive immune response starting with a macrophage encountering a pathogen and encompassing both the cellular (T-cell) and humoral (B-cell) responses.

Physiology of the Upper GI Tract

"What does the stomach do and how is it controlled?"

Model 1: The Swallowing Reflex

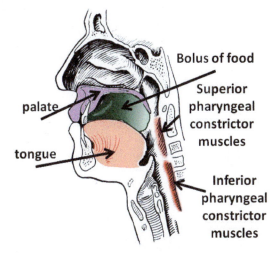

Bolus of food

Superior pharyngeal constrictor muscles

palate

tongue

Inferior pharyngeal constrictor muscles

(a) Tongue forms bolus and moves to pharynx

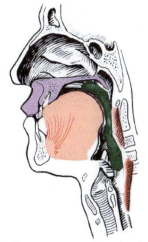

(b) Soft palate is elevated, inferior constrictor muscles relax, epiglottis closes

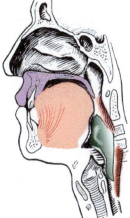

(c) Superior constrictor muscles contract, forcing bolus into esophagus

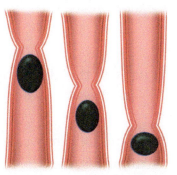

(d) Peristalsis of esophagus delivers bolus of food to stomach

Copyright © 2015

Critical Thinking Questions

1.   Part (a) of the model shows **Phase 1** of the swallowing reflex. What muscular organ is responsible for forming the **bolus** or ball of food and moving it into the pharynx?

    a. Is this organ made of cardiac, smooth, or skeletal muscle?

    b. Is this muscle type voluntary or involuntary?

    c. Is Phase 1 of the swallowing reflex voluntary or involuntary?

2.   Parts (b) and (c) comprise **Phase 2** of the swallowing reflex. What structure in the pharynx closes off the trachea and prevents choking?

    a. In the first Part of Phase 2, do the inferior pharyngeal constrictor muscle contract or relax?

    b. What will this do to the esophagus?

    c. After the food moves into the upper esophagus, what do the superior constrictor muscles do?

    d. What will this cause the bolus of food to do?

3.   Part (d) of the model shows **Phase 3** of the swallowing reflex. What type of specialized muscle contraction moves food through the esophagus and to the stomach?

    a. Can you stop swallowing once you've started? [try it if you need to]

    b. Based on this, are Phases 2 and 3 voluntary or involuntary?

Application

4.   You are examining the histology of the esophagus and pharynx. You notice a great deal of skeletal muscle in your slide. Explain which end of the GI tract your are examining and how you know this based solely on the presence of skeletal muscle.

Model 2: Components of Gastric Juice [memorize this!]

Component	Source	Function
Pepsinogen	Chief cells of gastric glands	Precursor to pepsin
Pepsin	Made from pepsinogen in the presence of HCl	Protein-splitting enzyme that breaks down nearly all dietary protein
Hydrochloric acid	Parietal cells of gastric glands	Denatures proteins and causes pepsinogen to be converted to pepsin
Mucus	Goblet cells, mucous neck cells (glands)	Forms a thick alkaline protective layer over the gastric mucosa (stomach wall)
Intrinsic factor	Parietal cells of gastric glands	Essential for the absorption of vitamin B_{12}
Gastrin (hormone)	G-cells of gastric glands	Stimulates gastric glands to release more gastric juice, makes stomach churn

Critical Thinking Questions

5.   Fill in the following concept map:

_____ →secrete _____ → is converted to _____

      (cells)   (precursor)   (active enzyme)

6.   What two substances are secreted by the parietal cells of the gastric glands?

      a. Write the functions of the substances next to their name.

7.   What provides the protective barrier that keeps you from digesting your own stomach?

       a. Where is it produced?

Application

8.   A deficiency of one of the components listed above can result in a gastric ulcer, or erosion. Which substance is it?

9.   A deficiency of another gastric juice component can result in anemia, which one?

Model 3: Regulation of Gastric Secretion

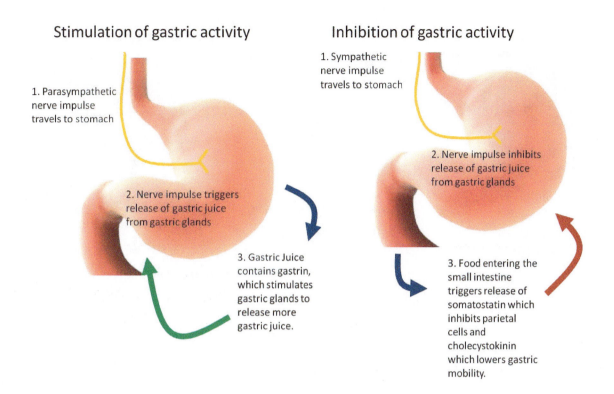

Stimulation of gastric activity

1. Parasympathetic nerve impulse travels to stomach

2. Nerve impulse triggers release of gastric juice from gastric glands

3. Gastric Juice contains gastrin, which stimulates gastric glands to release more gastric juice.

Inhibition of gastric activity

1. Sympathetic nerve impulse travels to stomach

2. Nerve impulse inhibits release of gastric juice from gastric glands

3. Food entering the small intestine triggers release of somatostatin which inhibits parietal cells and cholecystokinin which lowers gastric mobility.

10.   Which branch of the autonomic nervous system will stimulate gastric secretion and mobility?

a. Which branch will inhibit gastric secretion and mobility?

11.   According to the model, there is one hormone that stimulates gastric activity and two hormones that inhibit gastric activity. In the space below list those hormones and their effects on the stomach.

Model 4: The Three Phases of Gastric Secretion

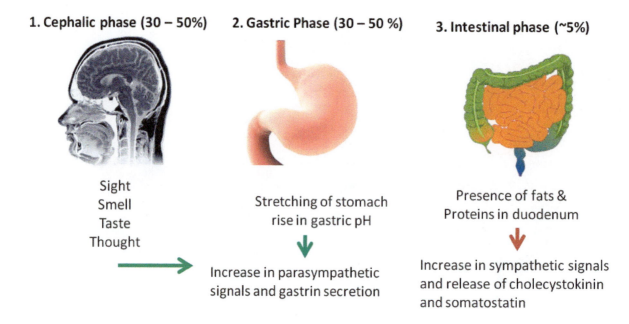

1. Cephalic phase (30 – 50%)   2. Gastric Phase (30 – 50 %)   3. Intestinal phase (~5%)

Sight
Smell
Taste
Thought

Stretching of stomach
rise in gastric pH

↓

Increase in parasympathetic
signals and gastrin secretion

Presence of fats &
Proteins in duodenum

↓

Increase in sympathetic signals
and release of cholecystokinin
and somatostatin

Critical Thinking Questions

12.   When your tummy rumbles and your mouth waters because you are thinking about your favorite food, which Phase of gastric secretion is being activated?

   a. What percentage of total gastric secretion can be accounted for by this Phase?

13.   What are the two triggers that initiate the gastric Phase of gastric secretion?

   a. What does your group think will cause the gastric pH to rise?

   b. What percentage of total gastric secretion can be accounted for by this Phase?

14.   Does the intestinal Phase stimulate gastric secretion or inhibit it? Hint: think about what sympathetic signaling and those two hormones do]

Application

15. Based on what you know about how food enters the stomach and how gastric secretion is regulated, answer the following as a group: Why is it easier to overeat if you eat really fast?

Anatomy of Ventilation

"What does it take to draw breath?"

Model 1: The Thoracic Cavity Before and During Inspiration

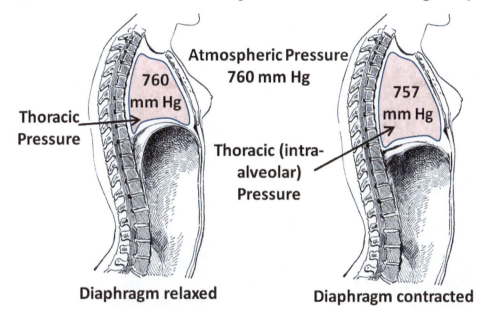

Diaphragm relaxed   Diaphragm contracted

Critical Thinking Questions

1.   According to the model, what is atmospheric pressure outside the body?

     a. What is the pressure of the thoracic cavity with the diaphragm at rest

2.   What should be the net direction of air flow [into the lungs, out of the lungs, or no flow] with the diaphragm at rest?

3.   Look at the picture on the right. Does the thoracic cavity appear larger or smaller with the diaphragm contracted?

     a. What effect does this change in thoracic volume have on the pressure inside the alveoli of the lungs?

     b. What should be the net direction of air flow with the diaphragm contracted?

POGIL
WWW.POGIL.ORG
Copyright © 2015

Application

4.   Be able to defend or explain the following statement: "The driving force behind pulmonary ventilation is atmospheric pressure"

5.   Explain why aircraft traveling at very high altitudes (atmospheric pressure decreases with altitude) must have pressurized cabins.

Model 2: Maximal Inspiration

Tidal (normal) inspiration   Maximal inspiration

Sternocleidomastoid muscle moves sternum up and out

External intercostal muscles pull ribs up and out

Diaphragm contracts

Pectoralis minor further elevates ribs

Diaphragm contracts more

Critical Thinking Questions

6.   Look at Model 2. What muscles besides the diaphragm are playing a part in a tidal inspiration?

7.   According to the model, what is the anatomical result of contraction of the external intercostal muscles?

        a. How will this affect the pressure in the thoracic cavity?

8.   Looking at Model 2, what two additional muscles play a role in <u>maximal</u> inspiration?

9.   What is the anatomical result of the contraction of these two muscles?

10.   How will contraction of these muscles affect the intra-alveolar pressure?

Application

11.   What would you expect the intra-alveolar pressure to be (ball-park guess) at the end of a maximal inspiration?

Model 3: Mechanisms of Expiration

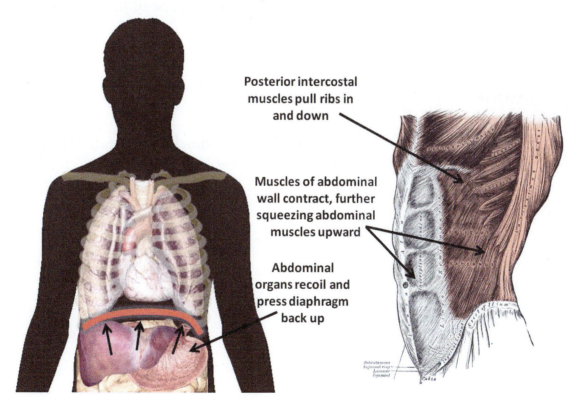

Posterior intercostal muscles pull ribs in and down

Muscles of abdominal wall contract, further squeezing abdominal muscles upward

Abdominal organs recoil and press diaphragm back up

Critical Thinking Questions

12. Look at the figure on the left. What role do your abdominal organs play in expiration?

13. How is this action aided by the muscles of the anterior abdominal wall?

    a. What does this do to the volume of the thoracic cavity?

    b. How will this change the pressure in the thoracic cavity?

14. What other muscles play a role in decreasing the volume of the thoracic cavity?

Application

15. Explain why getting punched in the abdomen knocks the breath out of you.

Model 4: Pneumothorax and Cross Section through the Thorax

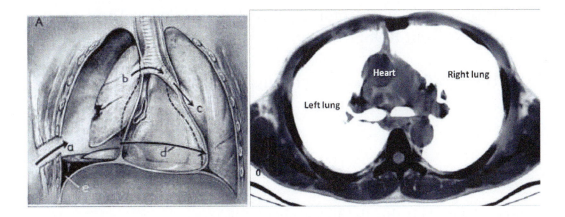

Critical Thinking Questions

16. Look at the left-hand picture above - there is a puncture in the right thoracic wall. As the diaphragm contracts and the volume of the thoracic cavity increases, will air enter the thoracic cavity through the trachea, the puncture or both?

   a. "Pneumo" is a prefix meaning "air" (for instance pneumatic tires are air-filled). Knowing this, explain why this condition is called a pneumothorax.

17. Looking at the picture on the right, does it appear that the right and left lungs share a common compartment?

18. Think way back - what do we call the structure that separates the right and left sides of the thoracic cavity?

Application

19. Explain why you can have one lung collapse and not the other.

Exercises (homework)

Respiratory distress syndrome is common in premature infants. Answer the following questions about RDS:

1.   What complex fluid is missing in infants with RDS?

2.   From which cells is this fluid secreted?

3.   Surfactant isn't usually secreted until 1-2 days pre-partum. Explain why RDS is always a factor in premature infants.

4.   How is RDS similar to a bilateral pneumothorax?

The Renin - Angiotensin System

"How do the kidneys regulate blood pressure?"

Model 1: The Renin - Angiotensin System

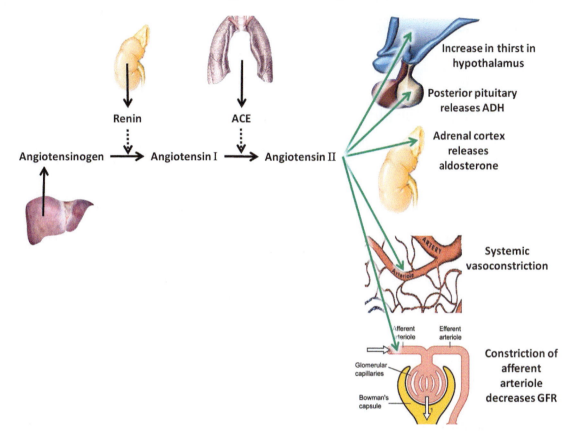

Critical Thinking Questions

1.   Based on the model, where does renin originate?

2.   Renin is an <u>enzyme</u>. Based on the model what is the function of the enzyme renin?

3.   The substrate for renin is the protein angiotensinogen. What organ secretes angiotensinogen?

     a. In what liquid tissue do you suppose angiotensinogen is converted to angiotensin I?

POGIL
WWW.POGIL.ORG
Copyright © 2015

4.   Of the two sites that secrete ACE, the <u>pulmonary epithelium</u> is the most important.

    a. What is the function of the *enzyme* ACE?

    b. Given its function, what does ACE stand for?

5.   What (if any) are the physiological effects of Angiotensin I?

6.   Based on the model, list the five major effects of Angiotensin II.

7.   Angiontensin II causes systemic vasoconstriction. What effect will this have on blood pressure?

8.   Angiotensin II caused the renal tubules to retain Na and Cl. Finish this sentence; Water follows .....

    a. If the renal tubules cause the blood to retain more sodium and therefore more water, what will that do to blood volume?

    b. What will be the effect of this on blood pressure?

9.   Angiotensin II also causes the adrenal gland to release aldosterone. How does aldosterone affect blood pressure?

10.   Angiotensin II acts particularly on the glomerular arterioles. It causes both the afferent and efferent arteriole to constrict.

a. What will vasoconstriction of these arterioles do to systemic blood pressure?

b. What effect will vasoconstriction of the afferent arteriole have on glomerular perfusion (filtration rate)?

c. What affect will this have on BP? Why?

11.   Angiotensin II stimulates the posterior pituitary to release ADH. What effect does this have on the kidney?

a. What is the effect on systemic blood pressure?

Application

12.   Many of you have grandparents or parents who take drugs called ACE inhibitors for their hypertension. This drugs bind to ACE and prevent it from carrying out its intended function. Review your answers to questions 5-11 and explain why taking these drugs is an appropriate therapy for hypertensive (i.e. have high blood pressure) patients.

Exercise

13.   Draw a flowchart that illustrates the renin-angiotensin system. Include the major organs and the effects of angiotensin II. [You should use the back of this page]

Water Homeostasis

"How does the body keep water balance stable?"

Model 1: Distribution of Body Water

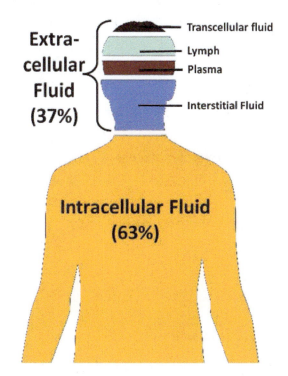

Critical Thinking Questions

1. Based on the model, what are the two major **compartments** where body fluids can be found?

2. Where is most of the body's fluid found?

3.   You should have a pretty good idea of what intracellular fluid, interstitial fluid, plasma, and lymph are. As a group, try to think of some places that you would find fluid that wouldn't fall into one of these 5 categories. Spokespeople: be ready to share your thoughts with the class.

4.   Think *way* back to cardiovascular physiology. What are the two forces that drive fluids between different compartments?

5.   What **ion** can the body move from one compartment to the other in order to manipulate osmosis? [Hint: water follows…..]

6.   There are two sources of water available to the human body.

a. What is the major way that people get water?

b. There is another, although less significant source of water called the **water of metabolism**. Librarians: look up aerobic respiration in your book or online. What is the byproduct of the electron transport chain's consumption of oxygen?

7.   There are two mechanisms by which the body excretes excess water, **excretion** and **evaporation**.

a. What are the two main forms of excreta that the body will use to eliminate water?

b. What is the main way the body loses water via evaporation?

c. OK, so sweating is easy, but there is another important way the body loses water via evaporation. You can figure it out by answering the following question: Why can you see your breathe in cold weather?

Got it? So: what, besides sweat, is a way the body loses water through evaporation?

8.   Obviously, diuresis is the major way the body regulates water balance. What hormone is going to be the principal mechanism by which your body regulates water balance?

Acid/Base Homeostasis

"How does the body maintain a stable pH?"

Model 1: Chemical Buffering Systems

A buffer is a compound that resists a shift in pH. When pH is falling (becoming more acidic) the buffer tends to take-up H+ ions. When pH is rising (becoming more basic) the buffer tends to donate H+ ions to the solution.

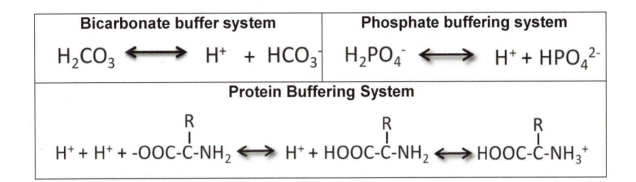

Bicarbonate buffer system	Phosphate buffering system
$H_2CO_3 \longleftrightarrow H^+ + HCO_3^-$	$H_2PO_4^- \longleftrightarrow H^+ + HPO_4^{2-}$

Protein Buffering System

$$H^+ + H^+ + {}^-OOC\text{-}\overset{\overset{\displaystyle R}{|}}{C}\text{-}NH_2 \longleftrightarrow H^+ + HOOC\text{-}\overset{\overset{\displaystyle R}{|}}{C}\text{-}NH_2 \longleftrightarrow HOOC\text{-}\overset{\overset{\displaystyle R}{|}}{C}\text{-}NH_3^+$$

Critical Thinking Questions

1.   If the pH of a system is rising, is it becoming more acidic or more basic?

2.   Does that mean it has more or less H+ ions?

3.   Based on this, will a buffer need to produce H+ ions or take them up?

4.   Look at the bicarbonate buffer system. If pH is rising will that favor the <u>formation</u> or <u>dissociation</u> of carbonic acid (H_2CO_3)?

5.   If pH is falling, will <u>formation</u> or <u>dissociation</u> of carbonic acid be favored?

POGIL
WWW.POGIL.ORG
Copyright © 2015

6.   Look at the phosphate buffering system. What will favor the formation of **dihydrogen phosphate** (H_2PO_4)? A drop in pH (or the solution becomes more acidic]

   a. What will favor the dissociation of dihydrogen phosphate?

7.   Look at the protein buffering system and circle the part of the model that will be the most acidic. Should circle the far left ion with two free H+

   a. Will a rise in pH favor the equation moving to the right or to the left?

   b. Will a drop in pH favor the equation moving to the right or to the left?

Model 2: Carbon Dioxide (CO_2) combines with Water (H_2O) in the blood to form Carbonic Acid (H_2CO_3) which is a source of H+ ions, and therefore tends to lower the pH of blood.

8.   Do you think you can lower the amount of water in the blood sufficiently to prevent the formation of carbonic acid?

9.   So, in order to prevent the formation of carbonic acid, what will your body have to eliminate from the blood?

10.   What organ will your body use to eliminate waste CO_2?

11.   As a group, be able to defend the following statement: The lungs play a vital role in blood pH homeostasis.

12.   The term <u>alkaline tide</u> refers to the removal, from the blood, of excess base left over from the production of stomach acid. What fluid actually becomes alkaline (basic) as base is removed from the blood?

13.   So, in addition to the lungs, what other organ is playing a role in maintaining blood (and therefore body) pH?

14.   Carbonic acid, dihydrogen phosphate, and proteins are always present in body fluids (both intra and extra-cellular). Based on this, would you expect these **chemical buffering systems** to act in seconds, minutes, or hours?

15.   **Physiological buffering systems** (like respiratory and renal clearance) tend to take longer. Both the lungs and the kidneys have many liters of blood passing through them an hour. However, greater surface area will allow for faster regulation of pH. Based on this which physiological buffering system is faster?

Meiosis

"How are genes sorted into gametes?"

Model 1: Mitosis Review

Mitosis is divided into 5 segments. The first **interphase** is usually considered the period between divisions, but it is also when duplication of the genome occurs. **Prophase** is the first phase of the actual mitotic division process. During prophase the DNA condenses into distinct visible **chromosomes** composed of one or two **chromatids** joined together by a single **centromere**. The next phase, **metaphase**, is characterized by the movement of the chromosomes to the center of the cell where they will line up with all the centromeres in a row. As soon as the chromosomes begin to separate at the centromere, **anaphase** has begun.

Demonstration: After you've had a chance to read the paragraph above, watch as the instructor models how chromosomes should be drawn.

Critical Thinking Questions (make sure you do this in pencil!!!)

1.   In the space below draw a circle to represent a cell. Our mythical cell is diploid, with 2 sets of chromosomes. Draw a **prophase** cell with two copies of chromosomes 1 (long) and two copies of chromosomes 2 (short), each with 2 chromatids (since gene duplication occurred prior to prophase).

2.   Now draw another circle. This time the cell is in **metaphase**, with all the chromosomes lined up along the mid-line of the cell. Draw **centromeres** at either pole of the cell with **spindle fibers** stretching to each centromere.

3.   Draw another circle. Draw this cell in **anaphase** with the chromosomes splitting at the centromere, so the each end of the cell will receive 2 long chromosomes and 2 short chromosomes, each with one chromatid.

4.   Now draw one more circle. This time the cell is in **telophase** with each end of the cell containing 2 copies of chromosome 1 (long) and two copies of chromosome 2 (short). Each chromosome should have 1 chromatid and 1 centromere.

Model 2: Meiosis I in Males

Meiosis is kind of like mitosis done twice. Consequently, it is divided into **meiosis 1** and **meiosis 2** with each part containing all the steps of mitosis. There are some differences though, so we will look at meosis I and compare it to mitosis.

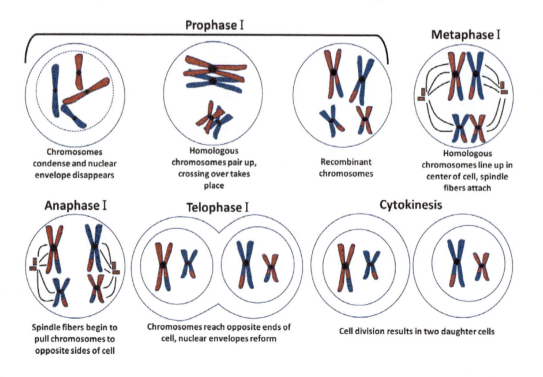

Critical Thinking Questions

5.   Observe how the chromosomes line up in prophase 1. How is this different from prophase in mitosis?

6.   Look very carefully at the chromosomes lined up in metaphase 1. Are they different than they were before they lined up?

a. What is different about them?

b. What event during prophase 1 can account for this change?

7.   Count the centromeres in the daughter cells. How many copies of each chromosome are there?

8.   How many of **each** chromosome did the parent cell we started with in Model 2 have?

a. Is this cell diploid or haploid?

b. How many chromosomes does each daughter cell at the end of meiosis 1 have?

c. Are these cells diploid or haploid?

d. Why is meiosis 1 referred to as the <u>reduction division</u>?

Model 3

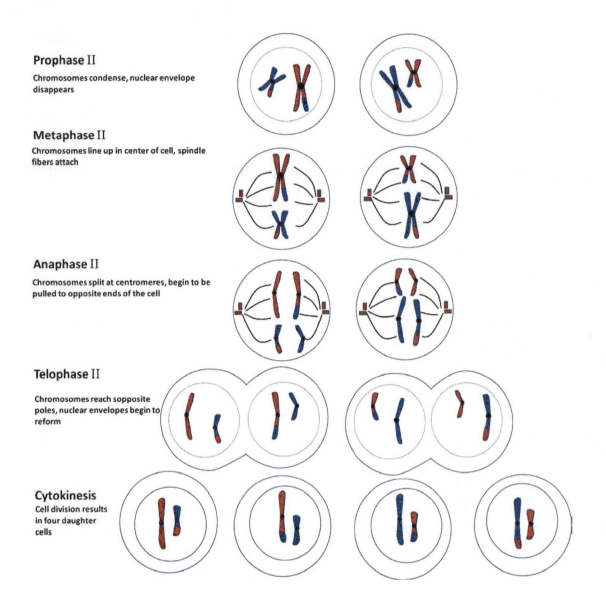

Prophase II
Chromosomes condense, nuclear envelope disappears

Metaphase II
Chromosomes line up in center of cell, spindle fibers attach

Anaphase II
Chromosomes split at centromeres, begin to be pulled to opposite ends of the cell

Telophase II
Chromosomes reach sopposite poles, nuclear envelopes begin to reform

Cytokinesis
Cell division results in four daughter cells

9.   Compare Models 2 and 3 - which is more like mitosis, meiosis 1 or 2?

10.   The cell we started with was diploid - it had two copies of each chromosome.

a. How many parents do you have?

b. Where did the two copies of each chromosome come from?

11.   Now look at the 4 daughter cells that resulted from meiosis. How many copies of each chromosome does each daughter cell have?

a. Look closely at each daughter cell. Are all of those chromosomes entirely paternal or maternal? [paternal chromosomes are blue, maternal chromosomes are red]

b. Based on your answers to the previous questions, what function does meiosis serve besides sorting chromosomes into gametes.

The Menstrual Cycle

"What is the relationship between sex hormones and the anatomy
of the female reproductive organs?"

Model 1: The Menstrual cycle

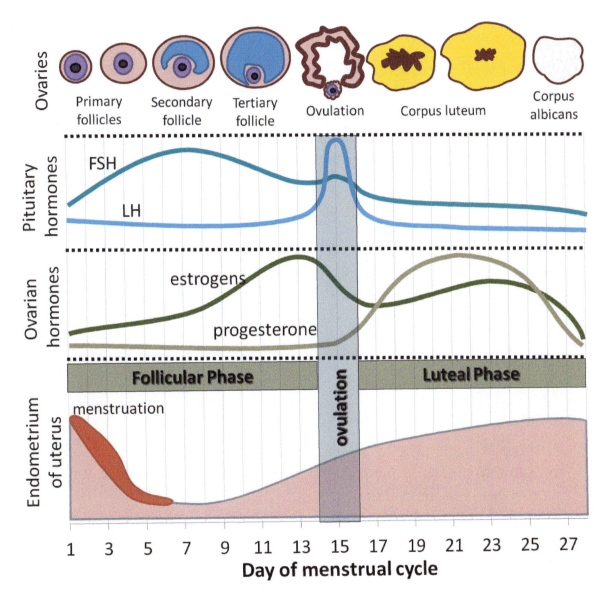

Critical Thinking Questions

1.   Look at the 'endometrium of the uterus' section of the model.

       a. Does it appear that there is any bleeding (as indicated by a red layer) on day 28 of the cycle?

       b. Does it appear the there is any bleeding on day 1 of the cycle?

2.   Based on your answers to Question 1, on what day does bleeding begin in the menstrual cycle?

       a. So when does the menstrual cycle begin?

3.   Based on the model, what is the proper term for the period of bleeding that occurs during the first several days of the menstrual cycle?

4.   According to the model, what is happening to the ovarian follicle during the first two weeks of the cycle?

5.   There are two large phases of the menstrual cycle, which one comprises approximately the first two weeks of the cycle?

       a. Why is it called that?

6.   Which sex hormone (meaning produced by the ovaries) is produced in the highest abundance during the first half of the cycle?

a. Does there appear to be any kind of relationship between the size of the ovarian follicle and the amount of this hormone? If so, what is the relationship?

b. Based on your answer to (a), where does estrogen come from?

c. If the <u>follicle</u> is producing estrogen, which gonadotropin (produced by the pituitary gland) is likely <u>stimulating</u> the production of this sex hormone?

7.   What is the relationship between FSH, the ovarian follicle, and estrogen?

8.   According to the model, what event separates the follicular phase from the luteal phase?

9.   Does it appear that any hormone(s) might be associated with this event? Which one(s)?

a. Which ovarian hormone reaches its peak concentration first?

10.   Based on the timing of hormone release in the model, fill in the blanks in the following sentence:

_____ peaks causing the release of _____ which triggers _____.
    (hormone)   (hormone)   (event)

11.   According to the model, what is the post-ovulatory phase of the menstrual cycle called?

12.   Which ovarian structure is developing and degenerating during this phase?

    a. Where did this structure come from?

13.   Which hormone's concentration appears to mirror the development and degeneration of this structure?

14.   What is happening to the thickness of the endometrium during this phase?

15.   As far as you are aware, does the uterus produce any hormones? (have we ever mentioned the uterus as an endocrine gland?)

16.   As a group, compose a grammatically correct complete sentence (in English) that describes the relationship between the corpus luteum, progesterone, and the endometrium.

17.   When are ovarian hormones at their lowest concentrations?

    a. What is the relationship between the levels of ovarian hormones and the ending of one cycle and beginning another?

Application

18.   Explain how measuring the ratio of estrogen to progesterone could potentially tell you where a woman is in this cycle.

19.   As a group, explain why maintaining high levels of progesterone would prevent ovulation.

Making a Person: From Zygote to Gastrula

Model 1: Earliest Stages of Embryogenesis

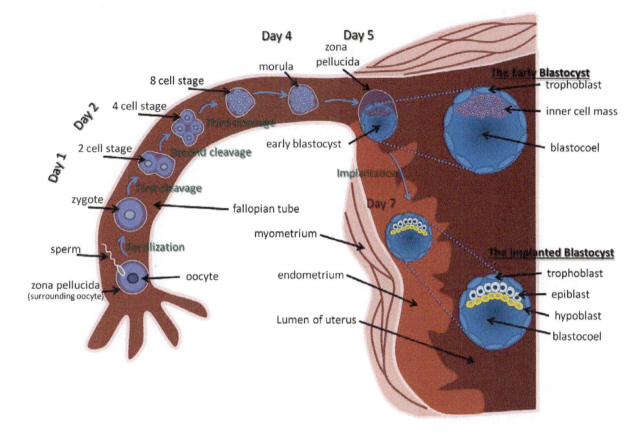

Critical Thinking Questions

1. According to the model, what is the name of the cell that results from the fertilization of the egg by the sperm?

    a. What is the name of the process that produces this cell?

2. What stage occurs after the first cleavage? After the second cleavage?

    a. Based on your answers above, what does the term 'cleavage' mean in this case?

3.   What is the **embryo** called on day four when it has become a solid ball of cells?

4.   On Day 5, the **early blastocyst** hatches. From what structure does it hatch (i.e. what does the blastocyst emerge from)?

5.   What is the name of the single layer of cells that completely surrounds the blastocyst?

      a. What are the other two components of the blastocyst?

      b. Someone in your group look up the definition of the biological term 'coel'. After they have explained it to the group, write a consensus definition of the **blastocoel** in the space below.

6.   What major event takes place between Days 5 and 7?

7.   Following this event the **inner cell mass** differentiates into a **bilaminar disk**. What are the two layers (lamina) of the bilaminar disk?

Model 2: The second Week of Life

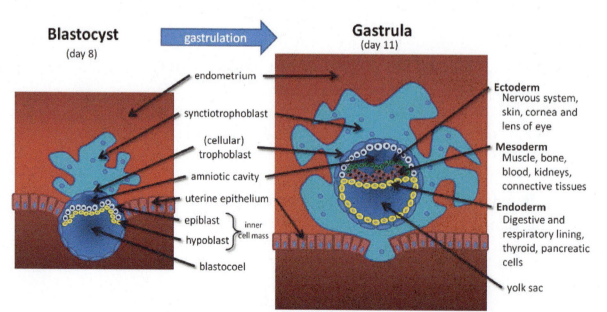

Critical Thinking Questions

8.   According to the model, what structure grows from the trophoblast into the surrounding uterine (endometrial) tissue?

      a. What cavity is beginning to open up between the trophoblast and the inner cell mass at this time?

9.   By Day 11, there are two spaces inside the embryo, one cavity and one sac. List those two spaces in the space below.

      a. In addition to the formation of these two cavities, another important event occurs during **gastrulation**. In the space below, list the three **germ layers** that are formed during gastrulation and the tissues of the body that are derived from these three layers.

10.   All of the body's tissues and organs derive from these three germ layers. What structure in the <u>early blastocyst</u> gave rise to these three germ layers?

      a. Do the trophoblast, amnion, or yolk sac appear to develop into any tissues that will be found in the child at birth?

      b. Why are these called <u>extraembryonic tissues</u>?

Application

11.   Cardiac and renal defects often occur together in newborns, as do skin and nervous disorders. As a group, explain why these (and other) birth defects often occur together.

12.   As a group discuss the consequences of defects occurring in these early stages of development. Do you think spontaneous abortion (miscarriage) is more likely during this time or later in pregnancy? Be able to explain **why**.

Matching

_____1. Zygote	A. Structure in blastocyst that eventually gives rise to all organs and tissues
_____ 2. Cleavage	B. Layer of the embryonic disc closest to the yolk sac
_____ 3. Morula	C. Solid ball of cells resulting from early cleavage events
_____ 4. Blastocyst	D. Embryonic stage that implants into endometrium
_____ 5. Inner cell mass	E. the diploid cell that results from fusion of egg and sperm
_____ 6. Epiblast	F. Comprised of the three germ layers in gastrula, it is the only part that will contribute toward tissue of the baby
_____ 7. Hypoblast	G. Layer of cells that surrounds the blastocyst and will eventually give rise to the placenta
_____ 8. Trophoblast	H. Cell division without growth, occurs early in embryogenesis
_____ 9. Germ layer	I. One of three primitive layers in gastrula that give rise to all tissue and organs
_____ 10. Embryonic disc	J. Layer of embryonic disc closest to the amniotic cavity

Listing

In the space below, place the following in the proper order.

Blastocyst

Fertilization

Formation of amniotic cavity

Four-cell stage

Gastrulation

Hatching

Implantation

Morula

Two-cell stage

Zygote

Mendelian Genetics

Model: Mini-Lecture on Mendelian Genetics

Problems

1.   Autosomal recessive conditions are carried on one of the 22 autosomal chromosomes. In order for a person to express the attribute (or have the condition), they must possess two recessive alleles. People who are heterozygous for the attribute are said to be **carriers** of the condition because they don't have it, but they can pass it on to their offspring.

     Cystic fibrosis is an autosomal recessive disease. If the alleles are written as CF (dominant) and cf (recessive) perform the following crosses.

     a. Two carriers have a child - write out this cross below.

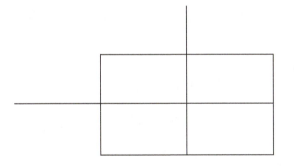

     b. What is the probability that a child will have cystic fibrosis?

     c. What is the probability that a child will be a carrier?

POGIL
WWW.POGIL.ORG
Copyright © 2015

2.   Autosomal dominant conditions require only a single copy of the dominant allele for the person to express the characteristic (or have the condition). Only homozygous recessive individuals do not have the condition.

Huntington's disease is an autosomal dominant condition. If the alleles are written as H (dominant) and h (recessive), perform the following crosses.

a. Someone who is heterozygous for the Huntington's allele has a child with a non-Huntington's person. Write out this cross below.

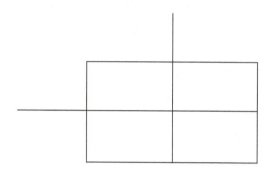

b. What is the probability that their child will have Huntington's disease?

c. What is the probability that their child will be a carrier?

3. Incomplete dominance is different than regular dominant or recessive inheritance. Loci that exhibit incomplete dominance follow the following pattern: Homozygous dominant individuals are unaffected, heterozygous individuals have a mild form of the condition (called the trait), while homozygous recessive individuals have the fully-severe form of the condition.

Sickle cell anemia is a condition that exhibits incomplete dominance. HB is the dominant allele while hb is recessive. Perform the following cross.

a. Someone with the sickle cell <u>trait</u> has a child with someone who has full-blown sickle cell anemia. Write out the cross below.

b. What is the probability that a child will have sickle cell anemia?

c. What is the probability that a child will have the sickle cell trait?

d. What is the probability that a child will be unaffected?

More Complex Forms of Inheritance
"What about loci that don't follow regular dominant/recessive patterns?"

Model 1: Sex-linked inheritance

Sex-linked inheritance refers to conditions carried on the X chromosome. Sex-linked traits, when present in a male (XY) are always expressed. However in women, they behave like regular recessive conditions (e.g. the woman must have two copies of the allele for the condition in order to express it).

An example of a sex-linked trait is hemophilia. A hemophiliac man will have the genotype $X^h Y$, whereas for a woman to be a hemophiliac, she must have the genotype $X^h X^h$. Women with the phenotypes $X^H X^H$ or $X^H X^h$ show no symptoms and are completely healthy. The only healthy genotype for men is $X^H Y$ Knowing this, answer the following questions:

Critical Thinking Questions

1. Women can be carriers of sex-linked conditions. Write the genotype of a woman who is a carrier of hemophilia:

2. Write out the genotype of a man who is not himself a hemophiliac:

3. In the space below, write out a Punnett square showing a cross between a woman who is a carrier of hemophilia and a man who does not have the condition.

POGIL
WWW.POGIL.ORG
Copyright © 2015

4.   What is the probability that this couple will produce a **daughter** that does not carry the hemophilia allele?

5.   What is the probability that this couple will have a **son** with hemophilia?

6.   What is the probability that this couple will have a son who does not have hemophilia?

7.   What is the probability that this couple will have a daughter that is a carrier of hemophilia?

8.   If a sonogram shows the couple is having a boy, what is the probability that he will be a hemophiliac?

Model 2: Co-dominant Inheritance

Co-dominant alleles are both expressed if they are present in the same individual. The ABO blood group gene is an excellent example of this. Both the A and B alleles are dominant, while the O allele is recessive.

Using this information, answer the following questions:

9.   What is the genotype of someone with type AB blood?

10.   What are the possible genotypes of someone with type A blood?

    a. Type B blood?

11.   What is the genotype of someone with type O blood?

12.   Dr. Brown has type A blood. His father has type O blood.

      a. What is the only allele that Dr. Brown's father could donate to him?

      b. What must Dr. Brown's genotype be?

13.   Dr. Mrs. Brown has type O blood. What are the chances that our new baby
      will have type A blood?

14.   What are the chances that the new baby will have type O blood?

Model 3: Polygenic Characteristics

Polygenic characteristics or conditions are caused by more than one gene (locus).
Examples include eye and skin color. Since we are dealing with more than one
locus, look up to the board and work together to answer the following questions.

15.   Blood type is actually determined by two genes, the ABO gene and the Rh
      factor gene. Dr. Brown has A negative blood. The negative allele (-) is
      recessive and the positive allele (+) is dominant. What is Dr. Brown's
      genotype for Rh?

      a. What is his full blood genotype (ABO and Rh next to each other)?

16.   Dr. Mrs. Brown has Rh+ blood, but her father is Rh-. What must her full
      blood type genotype be?

17.   When gametogenesis occurs each gamete Dr. Brown produces will have one allele from each locus. What are the possible gametic combinations that Dr. Brown can produce?

18.   What are the possible gametic combinations that Dr. Mrs. Brown can produce?

19.   In the space below write out the cross between Dr. Mr. and Dr. Mrs. Brown.

20.   What is the probability that they will have a baby with O positive blood?

    a. What is the probability that they will have a baby with O negative blood?

    b. What is the probability that they will have a baby with A positive blood?

    c. What is the probability that they will have a baby with A negative blood?

Model 20D

Eye color is another polygenic characteristic. There are two loci that control the color of the pigment cells in the iris of the eye (this is the main, but not the only thing that determines eye color). The green 'G' allele and blue 'B' allele are the two that determine the levels of pigment in the iris according to the table below.

Genotype	Phenotype	Phenotype
GGBB	Dark Brown/Black	
GGBb	Brown	
GGbb	Medium Brown	
GgBB	Light Brown	
GgBb	Hazel	
Ggbb	Green	
ggBB	Blue/Grey	
ggBb	Dark Blue	
ggbb	Light Blue	

21. Write out a dihybrid cross between two parents with Hazel eyes and calculate the probabilities of a child having each possible eye color.

Genotype	Phenotype	Phenotype	Probability
GGBB	Dark Brown/Black		
GGBb	Brown		
GGbb	Medium Brown		
GgBB	Light Brown		
GgBb	Hazel		
Ggbb	Green		
ggBB	Blue/Grey		
ggBb	Dark Blue		
ggbb	Light Blue		

Image Credits for
Anatomy and Physiology: A Guided Inquiry

Title Page
Mouse_sonya. *Stylish Skeleton Photo* on Dreamstime.com.

Articulations
Questions 4 – 9 (Forearm and Rib Cage):
Human Skeleton by Mikael Häggström, Public Domain

Joint Movements
Question 10:
Helland, Frode Inge. 1991. Silje Studio 11år, Creative Common Attribution, (Ballet Dancer).
Schertzer, Fanny. 2010. Hinano Eto There Where She Loved – Prix de Lausanne, Gnu Free Documentation License, (Contemporary Dancer).

Anatomy of Pulmonary Ventilation
Models 1 & 2:
Modifications of figures from the 1918 Edition of Gray's Anatomy, Public Domain.

Model 3:
Modification from the 1918 Edition of Gray's Anatomy, Public Domain, (Right Figure).
Häggström, Mikael. *Human Body Project* on Wikimedia commons, Public Domain, (Left Figure).

Model 4:
Brewer III, Lyman A., M.D., and Thomas H. Burford, M.D. *Surgery in World War II: Thoracic Surgery*, Volume II (1965). Medical department, United States Army, Public Domain, (Left Figure)
Heilman , James, M.D. Right-Sided Pneumothorax (right side of image) on CT Scan of the Chest with Chest Tube in Place, Creative Commons Attribution-Share Alike 2.5 Generic License, (Right Figure).

Cardiac Cycle 1
Model 1:
Ruiz, Mariana. *Blue Baby Syndrome*, Public Domain.

Cardiac Cycle 2
Model 1:
Pierce, Eric. Labeled Diagram of the Human Heart, Creative Commons Attribution-Share Alike 3.0 Generic license.

Endocrine Glands and Hormones
Model 2:
Nguyen, Marie-Lan. *Venus de Milo Statue*, Louvre, Public Domain.

Hemostasis
Question 8:
Wolberg, Alisa. *False Color SEM of a Blood Clot.* © Used with Permission.

Innate Immunity
Model 1:
Derived from Unattributed Works in the Public Domain, Modified by the Author.

Physiology of the Upper GI Tract
Model 1, Parts 1-3:
Modification from the 1918 Edition of Gray's Anatomy, Public Domain.

Model 4:
User Deradrian. Flickr. Creative Commons Share-alike 2.0,
        Generic License, (MRI of Brain).
Häggström, Mikael. *Human Body Project* on Wikimedia commons, Public Domain,
        (Intestines).

Receptors
Model 3:
Modification from the 1918 Edition of Gray's Anatomy, Public Domain.
        (Lamellar Corpuscle).

Skin and Tmperature Regulation
Model 2:
de Souza Telles, Daniel. Skin Diagram, Gnu Free Documentation License.
Lawrence, Micah. The Noun Project, Gnu Free Documentation License, (Running icon).

Sliding Filament Theory
Question 13:
Herring, Albert. 2012. Strawberry Festival Tug o War, 2012,
        Creative Common Attribution.

The Language of Science and Medicine
Model 3:
Vollmer, von Wilhelm. *Zeus*, Dr. Vollmer's Wörterbuch der Mythologie aller Völker,
3rd Edition Stuttgart 1874, Public Domain.
Model 4:
Bertram Mackennal, Edward. *Circe*: Statue, Public Domain.